Disasters in World H

Disasters in World History surveys the development of disaster studies as a discipline as well as presenting historical case studies and theories used by historians to understand disasters. Disasters, here defined as the complex interaction between natural hazards and specific human vulnerabilities, have frequently left a mark on human history. Cataclysms have toppled dynasties, fueled massacres, and shaped the culture of societies frequently affected by natural hazards. This volume fosters understanding of such events by considering both social science theory and the natural science concepts relevant to disaster studies. In addition, the text makes heavy use of an emerging psychological theory relevant to disaster studies: the behavioral immune system, which helps to explain why xenophobic behavior and even violence often erupt in the aftermath of disasters. Chapters consider specific examples of disasters: earthquakes, tsunamis, volcanic eruptions, climate change (including modern anthropogenic climate change or global warming), and tropical cyclones.

This book is an accessible resource, ideal for undergraduates and instructors in world history, environmental history, and disaster studies courses.

Benjamin Reilly is a Teaching Professor of History and one of the founding faculty members of Carnegie Mellon University's campus in Qatar. An environmental historian, Dr. Reilly is particularly interested in how humans interact with natural processes, especially disasters and infectious diseases.

Themes in World History
Series editors: Peter N. Stearns and Jane Hooper

The *Themes in World History* series offers focused treatment of a range of human experiences and institutions in the world history context. The purpose is to provide serious, if brief, discussions of important topics as additions to textbook coverage and document collections. The treatments will allow students to probe particular facets of the human story in greater depth than textbook coverage allows, and to gain a fuller sense of historians' analytical methods and debates in the process. Each topic is handled over time – allowing discussions of changes and continuities. Each topic is assessed in terms of a range of different societies and religions – allowing comparisons of relevant similarities and differences. Each book in the series helps readers deal with world history in action, evaluating global contexts as they work through some of the key components of human society and human life.

The Turkic Peoples in World History
Joo-Yup Lee

Punishment in World History
Peter N. Stearns

The Environment in World History (Second Edition)
Stephen Mosley

Globalization in World History (Fourth Edition)
Peter N. Stearns

The Atlantic Slave Trade in World History (Second Edition)
Jeremy Black

Disasters in World History
Benjamin Reilly

Disasters in World History

Benjamin Reilly

NEW YORK AND LONDON

Designed cover image: Candide, satire written by Voltaire in 1759. Illustration by Norman Tealby. "'This is the Last Day,' cried Candide." Lebrecht Authors / Lebrecht Music & Arts / Alamy Stock Photo.

First published 2025
by Routledge
605 Third Avenue, New York, NY 10158

and by Routledge
4 Park Square, Milton Park, Abingdon, Oxon OX14 4RN

Routledge is an imprint of the Taylor & Francis Group, an informa business

ISBN: 978-1-032-56695-5 (hbk)
ISBN: 978-1-032-56622-1 (pbk)
ISBN: 978-1-003-43680-5 (ebk)

DOI: 10.4324/9781003436805

Typeset in Times New Roman
by Taylor & Francis Books

Contents

Illustrations

Figures

Dedicated to Anita, Will, and Abel, the great loves of my life.

Acknowledgments

Many thanks to Supriya Sinha, Andrew Marshall in the main campus library, the CMUQ libraries staff, Allison Sambucini and Jane Hooper at Routledge, and to Peter Stearns. I am also grateful for the anonymous reviewers of the original book proposal, who provided useful suggestions that helped to shape the finished work.

This publication was made possible by the generous support of the Qatar Foundation through Carnegie Mellon University in Qatar's Seed Research program. The statements made herein are solely the responsibility of the author.

Introduction

What Is a Natural Disaster?

Natural disasters rarely arrive unannounced. Hours before a major quake, observers may report ghostly lights shining above the soon-to-be shaken landscape. Dead rivers and dried-up springs might start flowing again, miraculously, in the days before a volcanic eruption. On the evening before the arrival of a hurricane, the sky may take on an unearthly appearance, luminous and churning like a pot of boiling gold. And in the minutes before a tsunami, the sea may recede to a wondrous extent, revealing a temporary landscape of exposed crab traps, sunken ships, and thousands of flopping fish.

Although seemingly supernatural in their appearance, most of these phenomena can now be ascribed to well-understood natural causes. The reactivation of dead water courses before a volcano can usually be explained by the rising of the underground water table as magma heats it from below. Thin and rippled cirrocumulus clouds, which arrive before an approaching hurricane, can take on the appearance of molten metal when touched by a setting sun. A tsunami will cause the ocean to withdraw before it advances if the trough rather than the crest of the oscillating wave is the first part to touch ashore. As for earthquake lights, they are less understood, but may be photoelectrical effects powered by grinding movements in the tectonically-stressed portion of the Earth's crust.

Modern science, therefore, is better equipped than ever before to explain the natural hazards that lead to disasters. Nonetheless, a growing body of social science scholarship has argued convincingly that fully understanding any given natural disaster requires a deep appreciation, not just of the tectonic and meteorological forces involved, but also the local human context. Indeed, "natural" disasters are inherently human phenomena, during which underlying economic problems, social tensions, and political inequities can come into stark relief. As we will see repeatedly throughout this text, a full accounting of how world history has been impacted by disasters requires a cross-disciplinary approach that finds a middle ground between the natural and the social sciences.

DOI: 10.4324/9781003436805-1

Natural Disaster as a Theme in World History

As a result of their large scale and wide-ranging impacts, natural disasters events have an almost unmatched capacity to shape the trajectory of world history. Earthquakes, tsunamis, volcanic eruptions, climate change, and cyclones have helped to topple kings, contributed to the defeat of armies, and impacted cultural traditions. What is more, disasters often serve as moments of revelation, illustrating the functions (or dysfunctions) of human societies. As John Burnham (1988) has noted, disasters often serve to reveal "the makeup of the social fabric as people reacted under extreme stress." Anthropologist W. Lloyd Warner (1947) has made the same point: "there is much to learn about ourselves and the worlds we fashion," he quips, "when all hell breaks loose."

Disasters open useful windows into the past in other ways as well. Since the same hazards often strike quite different human societies, the study of natural disasters allows historians to make fruitful transnational comparisons. The impact of a disaster like an earthquake on any given society depends largely on local factors, such as settlements patterns, population density, architectural practices, socioeconomic structures, and religious beliefs. Similarly, the study of natural disasters can yield useful insights into how a single society has changed over the course of history. While natural hazards, such as the tsunami waves and cyclonic weather, are dependent on the laws of physics and thus change little over time, the societies they impact often do experience profound historical transformations over the centuries, changes that can reduce, alter, or even exacerbate the vulnerability of that society to natural hazards.

At the same time, historians of disaster must keep in mind that different disasters work on quite different temporal scales. Following the lead of influential French historian Fernand Braudel (1996), many historians distinguish between *histoire événementielle* (short-term events), the *longue durée* (long-term, recurrent historical cycles and structures), and the *conjuncture*, the zone of interaction between the two. As this book will illustrate, natural disasters operate on all three of these time scales. Earthquakes and many volcanic eruptions can have dramatic impacts on the *histoire événementielle*, leading to severe but localized and often temporary disruptions to society. Much the same is true of cyclones and tsunamis, though by their nature these events can impact a larger area. Nonetheless, although the havoc wreaked by these events is often temporary, it may have important long-term consequences. As we will see in Chapter 6, for example, cyclonic weather may have played an important role in contributing to the outbreak and in ensuring the success of the American War of Independence. Similarly, as we will discuss in Chapter 3, the Indian Ocean Tsunami of 2004 altered the trajectory of Indonesian history, interrupting an ongoing insurgency and creating a moment for reconciliation.

If disasters are recurrent, on the other hand, they may help to shape the underlying economic and cultural characteristics of a society and thus leave their mark upon that society's deeper historical structures. One way that disasters can impact the *longue durée* is by leading to the creation of a "culture of disaster" – the cultural framework by which a society comprehends the outbreak and mitigates the impacts of disasters. Seasonal cyclonic activity in places like the Caribbean and the Ganges Delta, for example, have played an important role in shaping local architecture, economic decisions, settlement patterns, and even social structures. In Japan, earthquakes and tsunamis influenced building styles, religious culture, and even political culture, since the occurrence and proper management of natural disaster were intimately tied to a regime's political legitimacy.

Multi-year and multiregional natural disasters such as climate change can exert an even more profound and lasting influence upon the historical record, in part because such disasters strike human society where we are most vulnerable: our rainfall and temperature-dependent agricultural systems. The sudden and sustained global cooling that accompanied the 536 CE climate event may have "resynchronized world history," in the words of David Keys (1999), playing a role in the Arab conquest of the Middle East, the collapse of the Mayan city of Tikal, and (more surprisingly) the legend of King Arthur. The interaction of repeated El Niño events and European imperial policies contributed to the "Late Victorian Holocausts" of the late nineteenth and early twentieth centuries. Longer-term cyclical climate change caused by decade- or century-long variations in solar forcing have had an even greater impact on history. Scholars have found the fingerprint of climate in the rise of Rome, the fall of Chinese dynasties, the flourishing of the Islamic Golden Age, and the turmoil of the seventeenth-century "Global Crisis."

In sum, natural disasters are well worth the attention of historians. For one thing, disasters matter; cyclones, climate change, and other disasters have all served as potent agents of change, both in the *événtemientelle* and the *longue durée.* Disastrous events also serve as "magnifying lenses," in the words of Van Bavel et al. (2020), "which expose aspects of past society that might otherwise have escaped the eye." Natural disasters are also useful to the historian as natural history experiments, allowing for side-by-side comparisons of societies far removed from each other both in space and time. Such comparisons allow scholars to tease out how different levels and types of vulnerability, in addition to different local economic, political, and cultural configurations, led to different manifestations of disaster. In keeping with this possibility, I have tried to pick case studies for the later chapters that represent both different parts of the world and different eras of world history.

But What IS a Natural Disaster?

The problem with studying natural disasters, however, is that scholars have profound disagreements on their definition, or even whether, properly speaking, they exist at all. As historian Jonathan Bergman (2008) points out, "disaster is one of those extraordinary phenomena that is ubiquitous yet indescribable." Different definitions of disaster are routinely used in different fields of scholarship and even by scholars within the same field. Citing the work of Enrico Quarantelli (1987), who tried in vain to find a universal definition of disaster, Ronald W. Perry (2015) has concluded "there is little basis in logic and little hope in practice that a single definition can be devised" for disasters "that is universally accepted and useful." What is more, even if an academic definition could be crafted, it would compete with popular conceptions of disaster that define them more broadly as major disruptions to human lives and livelihoods due to human error or natural forces. In keeping with this folk definition of the term, biologists Roger del Moral and Lawrence Walker (2007) admit that in their own work they use the term "disaster" rhetorically, when emphasizing "impacts that are widespread or severe," or even when they "simply want to focus [the] attention" of their reader!

Furthermore, the problem of defining disaster is complicated by politics, both outside of and within academia. "Natural disaster" is not a neutral term. As Ted Steinberg demonstrated forcefully in his 2000 book, *Acts of God*, American elites have repeatedly invoked the concept of natural disaster to deflect blame from themselves as well as to justify self-serving policies, such as coastal or floodplain development, that are "environmentally unsound, and socially, if not morally, bankrupt." In a similar vein, Sara Pritchard (2012) has noted that something seemingly apolitical like "pointing to environmental factors" to explain a disaster can be a political act if those factors are invoked to "diffuse, if not undermine, the responsibility and ultimate culpability of powerful groups" who had a hand in creating the catastrophe. One of the sharpest critiques of disaster as a discourse comes from the pen of Andy Horowitz (2020), who argues that "to name something a disaster is to decry its outcomes as illegitimate, and to call for restoration of the status quo, instead of suggesting that the status quo might have been illegitimate in the first place." As a result, some historians and other social scientists have become increasingly critical of the concept of "natural disaster" and tend today to emphasize the socioeconomic rather than natural roots of disasters. Human systems such as capitalism or neoliberalism, and not Mother Nature, are the real architects of disaster and the true villains of the tale. Rather than seeking to define the term "disaster" from the outset, then, it might be useful to survey the evolution of disaster studies as an academic discipline.

Disaster Studies: from Technocratic Relief to Postmodern Discourse

Interestingly, disaster studies was born in the shadow of the mushroom cloud: it arose as a discrete field of study during the Cold War era, a side effect of military and government-sponsored investigations into how people might behave in case of a nuclear war. Since the subject could not be studied directly, except perhaps in Hiroshima and Nagasaki, scholars turned to the study of non-nuclear disasters like cyclones and earthquakes as proxies for nuclear holocaust. At this stage, interest in disasters was primarily practical, focused on quantifying and mitigating risk to vulnerable populations. In keeping with such research, most scholars of the 1950s through the 1970s assumed that disasters were mainly the result of external, natural forces, such as seismic shocks and invading pathogens, and disaster therefore represented a dramatic departure from normal life. The solution to such problems was technocratic, to remake the human or natural landscape using modern engineering and medicine to reduce the level of risk. Ultimately, these scholars assumed, the goal of disaster mitigation was to get back to "normal," in other words, the pre-disaster state of affairs.

One side-effect of this framing of disasters is that historians were generally uninvolved in their study. According to John Burnham (1988), in the period before the mid-1960s, historians had written "virtually nothing" on natural disasters; rather, most literature on this subject was produced by natural scientists and public policy experts. Historians lacked a clear theory of disasters, or at least a concept of disaster that distinguished their thinking from the technocratic approach preferred by the science and public policy departments of the university. This began to change in the 1970s, however, with the growing popularity of the "vulnerability" approach to disaster proposed by geographers Phil O'Keefe, Ken Westgate, and Ben Wisner (1976). If disasters were primarily the result of extreme natural events, as the technocratic school believed, the number of natural disasters ought to be remaining steady over time or even decreasing, as our scientific and technological know-how improved. However, as O'Keefe and his colleagues pointed out, the reverse was occurring: large-scale disasters were becoming more and more common over the span of the twentieth century. Scholars in the technocratic camp who noted this increase in disasters ascribed it to increased human prosperity – Ian Burton, Robert Kates, and Gilbert White, for example, suggested in 1978 that disasters were increasing in frequency in large part because nowadays there was simply more people and more property in the world to be damaged. But as O'Keefe et al. pointed out, the twentieth century spike in disasters was much sharper in the relatively property-less developing world than in the wealthy developed West.

As a result, O'Keefe, Westgate, and Wisner proposed an alternative explanation for the apparent rise in modern-day natural disasters: catastrophes were becoming more frequent, not because large-scale natural hazards were becoming more common, but rather because these hazards were striking increasingly vulnerable human populations. Any study of disaster must therefore consider two factors: the natural hazard and the specific human social, economic, and political vulnerabilities of the affected population. In mathematical terms, this new understanding of "natural disasters" could be rendered: **Disaster = Hazard + Vulnerability**. And since historians were particularly skilled at using source materials to understand the social, economic, political, and other historical elements of a society that might lead to vulnerability, the field of disaster studies was now wide open to historical investigation.

The space for historians to study disaster was further widened by attacks on a core assumption of the technocratic interpretation of natural disasters: that disasters are events which begin with the outbreak of a natural hazard and end with recovery and a return to pre-disaster conditions. For some social scientists, nothing could be further from the truth. "The failure of a levee," historian Scott Gabriel Knowles (2020) contends, "like the first shot in a war, is the accumulation of political and material events that stretch back in time, often to indeterminate points." As a result, disasters are best understood not as events, but as long-term processes with deep historical roots. This concept has been taken to its logical extreme in Anthony Oliver-Smith's (2020) work on what he called Peru's "Five-Hundred-Year-Quake." While the triggering earthquake may have struck in 1970, Oliver-Smith contends, the vulnerability that turned this hazard into a disaster was far older, dating at least as far back as the Spanish conquest of Peru in 1532. Conquering conquistadors disregarded Indigenous Inca disaster mitigation techniques, such as dispersed settlement patterns and quake-resistant architecture. What is more, colonial Spain transformed the conquered population into an exploited underclass, and this situation changed little after Peruvian independence, in part because Peru remained in a disadvantageous position in an increasingly globalized world economy. As a result, Indigenous Peruvians at the time of the quake suffered from lack of health care opportunities, chronic poverty, and poor-quality housing built on potentially hazardous terrain, such as unstable floodplains or mountain slopes. By the time that Peru experienced a major quake on May 31, 1970, Oliver-Smith argues, these cumulative historical forces ensured that the society it struck "was in many ways already a catastrophe."

In recent years, Knowles (2020) has coined a new term for disasters with deep historical roots: "slow disasters," which are "not discrete events but long-term processes linked across time." While the unfolding of such disasters is often hard to discern at any given moment, Knowles contends, their cumulative impact may "claim much more life, health, and wealth across time than is generally calculated." One such disaster, for Knowles, is global warming, the roots of which reach down to the advent of the industrial age. Industrialization, in turn owes its own roots in the capitalist transformation of western societies in the early modern world, to the point that Knowles suggests we discard the term "Anthropocene," as the name of our current era of geological time is widely known, in favor of the "Capitalocene." By re-imagining our current predicament in terms of a capitalism-driven "slow disaster," Knowles argues, we may be able to hold the states and institutions contributing to these disasters accountable.

A second core belief of the technocratic approach that has come under assault from social scientists is the assumption that western top-down scientific and engineering interventions are a universal fix for natural disasters. As a result of this prevailing mindset about disasters, Kenneth Hewitt (1983) has argued, local knowledge and sensibilities are typically disregarded or even treated as part of the problem to be solved. Instead, western disaster management "experts" have typically treated non-western disaster victims as "pre-scientific" primitives, trapped in a desperate fight for survival against elemental forces. Given this imbalance of power and knowledge between expert and victim, Hewitt continues, the western technocrat feels empowered to speak "for" the people impacted by catastrophe, but "can find little value in dialogue *with* them or learning from them."

More recent scholarship has pushed this critique of the technocratic approach even farther. Since the beginning of the twenty-first century, the discipline of history has become increasingly interested in postmodern deconstructionism and discourse analysis, a change that some scholars have called field's "cultural turn." In disaster studies, this "cultural turn" has led to critical analysis of the dominant discourses surrounding natural disaster, including the idea of disaster itself. Greg Bankoff (2003), for example, has argued the West is hiding behind a discourse of "tropicality" – the idea that the developing world is inherently more subject to natural disasters due to geographic conditions – to avoid talking about the real reason why disasters are spiking in the Global South, namely globalization and the corrosive impact of international capitalism. In addition, "tropicality" gives the West the opportunity to play the part of the white savior to the benighted developing world, in the process, perpetuating the cultural hegemony of the West. Bankoff even argues that "vulnerability" is itself a rhetorical discourse as much as it is a tool of scholarly analysis. Vulnerability is not an abstract property that people have, but rather (according to Hilhorst and Bankoff, 2022) is the concrete result of historical socioeconomic exploitation. Like tropicality, Bankoff argues, the discourse of "vulnerability" dehumanizes non-western people, treating them as "inferior, untutored, and incapable victims" whose situation is their own fault and whose only hope is technocratic salvation from the West. Vulnerable populations, Bankoff argues, should combat their vulnerability with social protest and political action rather than blind adoption of western concepts of development and self-serving western technological and economic solutions, since "development" is itself a root cause of disaster.

Recent work in the social sciences has also cast doubt on a third core assumption of the technocratic school of disaster studies, that disasters are a departure from a "normal" state of affairs and that recovery from a disaster means a return to the pre-disaster conditions. Advocates of the vulnerability approach see natural disasters as arising not from extraordinary events but from the day-to-day conditions of life – they are, in the words of Kenneth Hewitt (1983), "*characteristic* rather than accidental features of the places and societies where they occur." Since disasters are triggered by human vulnerability, Hewitt and other scholars argue, and since poverty and underdevelopment are the most common causes of vulnerability worldwide, the concept of "recovery" is meaningless. After all, normalized poverty is what made the disaster possible in the first place.

For many scholars who have adopted the vulnerability approach, natural disasters and poverty are seen as inseparably intertwined. As Van Bavel et al. (2020) pointed out in a recent study:

> The poor tend to live in inherently hazard-prone locations, they lack the capital to invest in preventative measures or build up resource buffers for anything unexpected, they have more restricted access to helpful social networks, and frequently are disenfranchised from the political processes that can help steer policies more conducive to their protection.

Similarly, Anthony Oliver-Smith (2020) has noted that "the high correlation between disaster proneness, chronic malnutrition, low income, and famine potential" suggests that "the root cause of disasters" can mainly be found in the "structural imbalances between rich and poor countries."

Given the centrality of human vulnerability in causing "natural disasters," some scholars have suggested throwing out the term "natural disaster" entirely. Christian Pfister (2009), for example, has advocated swapping it for the phrase "nature-induced disaster," since this formulation "reflects the fact that such catastrophes are brought about by natural phenomenon without obscuring their anthropogenic [man-made] dimensions." Another group of disaster scholars want to go still farther and jettison, not just the term "natural," but "disaster" as well. According to Jacob Remes and Andy Horowitz (2021), who are pioneering the discipline of "critical disaster studies," "there is no such thing as disasters." Disasters are mere "interpretive fictions," they argue, and disasters can be spawned by the power of words alone without need for a real triggering event, natural or otherwise. The very idea of a disaster is itself a cultural construction; as Bankoff (2004) notes, cultural discourses not only play a role in defining the start of a disaster, they can even determine "what constitutes a disaster in the first place." For example, Bankoff points out the arbitrary convention by which hurricanes and cyclones are considered disasters, while traffic fatalities, which kill far more people on an annual basis, are not.

Advocates of critical disaster studies also follow Steinberg (2000) in suggesting that the concept of "natural disaster" has been hijacked by elites for their own purposes. Remes and Horowitz (2021) contend that declarations of disaster are best seen as rhetorical constructs designed by elites to return the society quickly to a pre-disaster state of affairs – a status quo that generally that favored the rich and the few while marginalizing the poor and the many. If this is accepted, then the job of a historian (according to Scott Gabriel Knowles, 2014) is to call out the organizations and people who are pushing for an "apolitical rendering of disaster" as a distraction for the real root cause of present-day disasters: the "risk-taking of late capitalism" in the modern neoliberal state.

The past seventy years, then, have witnessed a sea change in how disasters are understood by scholars. An initial approach prioritizing natural hazards has given way to a more balanced approach examining both hazard and vulnerability. This is now giving way in turn to a post-modernist critique of modern capitalist institutions and the very concept of disaster itself.

Newer Ideas: Risk Societies, Shock Doctrine, Resilience, and Cultures of Disaster

One concept that has gained considerable traction in disaster studies since it was introduced in the 1980s by Ulrich Beck is that of "risk societies" (Beck, 1992) – the idea that modern industrialized societies face a quite different set of disaster risks than the societies in the pre-industrial world. Unlike traditional societies, who were driven to disaster by short-term hazards arising from their interaction with the natural world, modern societies face risks that are global, chronic, and to some degree of their own making, manufactured by modernity itself, as epitomized by the nuclear melt-down of Chernobyl and the modern spike of cancer rates in the age of large-scale industrial pollution. Given the scale of these modern problems, followers of the risk society paradigm like Ryan Hagan (2021) find the notion of blaming human beings for disasters as "almost comically hubristic," since, at this point in our history, disasters are generally "the consequence of systems far beyond any individual or even institutional control." The same vast, impersonal systems that help to cause disasters also play a role in deciding who is victimized by a given disastrous event. While victims of pre-modern disasters tended to be those directly exposed to the event, modern disasters can disrupt or even claim the lives of people living far from the event if the disaster disrupts a technical system upon which people depend. Since modern humans are dependent on elaborate, globalized systems of food distribution, for example, a volcanic disaster may bring hunger to lands thousands of miles from the caldera of the volcano. Similarly, a cyclone can impact people living far outside of the reach of its winds by knocking out power grids, thus shutting down life-saving medical equipment. As a result of our modern dependence on technical systems for survival, Hagan contends, "deaths from infrastructure system failures," at least in the modern United States, "now routinely exceed fatalities directly caused by natural hazards" (ibid.).

The idea of the risk society, therefore, discounts the individual as an agent in understanding disaster. In contrast, the concept of the "Shock Doctrine" puts individual agency and intentionality in center stage: not in causing disasters, but in responding opportunistically to their aftermath. The term was coined in 2007 by social activist Naomi Klein, who argued for a pervasive tendency for elites to exploit the chaotic moments after a major disaster to push through ambitious social, economic, or political changes, since in the wake of catastrophe common people lack the attention or the resources to resist the innovations proposed by neoliberal elites (Klein, 2008). As we will see in the chapters that follow, a number of historians have found the "shock doctrine" idea useful in explaining the sweeping institutional changes that have been enacted in the wake of several recent major disasters, most notably Hurricane Katriana in 2005 and the Haitian Earthquake of 2010. However, Klein's work is controversial, and several scholars have accused her of oversimplified interpretations and misreading the historical record.

Another recent concept that has emerged in social science disaster studies is resilience, which is generally defined as the ability of an institution, system, or society to resist or accommodate to the impact of a major hazard and to preserve or restore the basic structures and functions that were disrupted by the disastrous event. To some degree "resilience" is a close cousin of the idea of vulnerability; while vulnerability may explain why a society crumbles in the face of a natural hazard, resilience may explain its ability to withstand the hazard quickly, efficiently, and with minimal changes. Several recent works on disaster studies have incorporated resilience into their titles and have attempted to assess the relative resilience of past societies in order to explain why a society did, or did not, experience a given disaster as a fundamental or existential crisis. However, like Klein's shock doctrine idea, the concept of resilience is controversial. D. E. Alexander (2013) notes that there is currently no scholarly consensus about resilience – there are in fact currently at least 12 definitions of resilience circulating in the academic literature – and he worries that in some recent scholarship "resilience is being used as little more than a fashionable buzz-word." I have my own reservations about the concept, which I will discuss below.

The idea of resilience is at least partially related to the last of the newer concepts this section will consider, that of "cultures of disaster." The term was first coined by Harry E. Moore in 1964 to describe the suite of cultural adaptations that societies have incorporated into their daily practices to cope with the ordinary and extraordinary hazards inherent in their lived environments. For example, Christian Pfister (2009) notes that European societies in Alpine areas deliberately retained natural forests above their mountain villages to reduce the threat of rockslides and constructed buildings with "wedge-shaped stonewalls" pointed upslope to divert avalanches. Similarly, the inhabitants of the Netherlands developed a toolkit of technical skills as well as legal structures to cope with the hazards inherent to living along the low-lying North Sea coast. Pfister also argues that natural disasters have also imprinted themselves on European religious culture, for example, through the regular invocation of prayers for protection against natural hazards and the construction of chapels in dangerous landscapes. These chapels serve a dual function, serving both as sacred sites for invoking supernatural protection against disaster as well as secular reminders of past hazardous events.

The "culture of disaster" that has been best studied, however, is that of the Philippines. As Greg Bankoff (2003) reminds us in his multiple works on the subject, the Philippines receives far more than its fair share of disasters; "the list of disasters in the Philippines," which includes earthquakes, volcanoes, and typhoons, "often reads like a record of Biblical plagues." Filipinos have adapted culturally to these recurrent threats in multiple ways, ranging from storm- and quake-resistant architectural designs to prevent house collapses to spatially fragmented agricultural land holdings and crop diversification to prevent total harvest loss in the case of a typhoon or flood. Bankoff even relates the Philippine Archipelago's unusually frequent natural disasters to aspects of the Filipino popular mindset, including their fatalistic attitude toward misfortune, their tendency toward black humor, their strong sense of community cohesion, and the almost universal belief in God, with only 1% of Filipinos identifying as atheist. As we will see in Chapter 1, however, some of the cultural adaptations that Bankoff describes may also reflect, not just local Philippine conditions, but also universal human psychology, which reacts to stressors in predictable ways.

A Personal Approach to Disaster Studies

So, at this point in the chapter, are we any closer to answering the question, "What is a natural disaster?" As is clear from the section above, academics are very far from scholarly consensus. Each of us must therefore stake out his or her own position, adopting approaches to the subject that align with our own convictions and the specific subject matter we are considering. Students will have the chance to refine their own views about natural disaster in the numerous case studies presented in the later chapters of the text, each of which is an opportunity to test abstract theories of disaster against real-world catastrophes.

However, since I am the one collecting, narrating, and to some degree interpreting these case studies, academic honesty requires me to reveal my own position for the reader. My preferred definition of natural disaster is the definition provided by Wisner et al. (2004) in *At Risk*, that disasters are disruptive events caused by "a complex mix of natural hazards and human action." I also ascribe to what Dorothea Hilhorst and Greg Bankoff (2022) call the "iron law" of disaster studies – that "disasters cannot be equated to the hazard but are outcomes of hazards encountering vulnerability mitigated by response capacities." That being said, I am also quite sympathetic to Pfister's (2009) point that the term "natural disasters" is misleading, potentially distracting the reader from the deep anthropogenic roots of all disaster events. His preferred term "nature-induced disasters" is an intellectually appealing solution to this problem, though it is too cumbersome and too little known to be put into routine use. Still, the reader should understand that whenever I use the term "natural disaster" in the text, I am using it in the sense of Pfister's "nature-induced disaster."

While I am convinced that the "vulnerability" approach to disaster is analytically useful, I am far less persuaded by its cousin concept, resilience. Vulnerability can be measured in fairly concrete ways – the number of feet below sea level, the number of kilometers from a volcano, the height of the levees, the percentage of people living below the poverty line, the percentage of coastal homes built to code, etc. On the other hand, resilience can only be measured after the fact, in terms of the disaster's outcome – resilience *is* therefore what resilience *does*. A population shows resilience by being resilient and recovering quickly, making the concept rather circular. I also think Bankoff is right to critique "resilience" on the grounds that it can be used by the powerful to duck their own responsibility for a calamity and shift blame to its victims. In the minds of elites, since resilience mitigates disasters, it logically follows that a population which has suffered a disaster must have lacked sufficient resilience. Hence, it has only itself to blame. "Resilience" discourse therefore runs the risk of taking us back to nineteenth-century moralistic critiques of poverty. Accordingly, I will make little use of the idea of resilience in the case studies below.

In contrast, I am generally persuaded by Oliver-Smith's "five-hundred-year" approach to disaster studies, a perspective that he employed effectively not only in his study of the Peruvian Earthquake of 1970 but also the more recent 2010 quake in Haiti. Oliver-Smith's approach illustrates what historians do best – we make events comprehensible by understanding them within their deeper sociopolitical, economic, and cultural contexts. That being said, the all-seeing historian must remember that disasters are usually understood quite differently by the people caught up within them. As we know from personal experience with the COVID-19 pandemic, most of us experience a major disaster as a traumatic shock and a sharp break from a recent "normal" past, a past that we hope to return to when the disaster finally ends. I am certainly sympathetic in theory to the idea that disasters are "processes" rather than events. However, we should keep in mind that the sweeping view of disasters we get from atop our academic ivory towers is quite different from the more myopic view of those trapped in an unfolding catastrophe.

Finally, I am only weakly persuaded by most works in the emerging historical field of "critical disaster studies." While I am sympathetic to some of the political positions to which these authors ascribe, and feel that these scholars can have clear insights into the current and future problems faced by modern society, the perspectives they give are much less relevant to the study of the historical past. Major disasters, after all, occurred long before the rise of capitalism, globalization, and the neoliberal state. I am also concerned that some practitioners of critical disaster studies, by reducing disaster to a discourse and denying that disasters even exist at all, are repeating one of the cardinal sins of the technocratic school of disaster studies: ignoring the ways in which events are actually experienced by people on the ground, in the eye of the hurricane, or the epicenter of the quake.

In addition, I dislike the tendency within postmodern studies which (in the words of David Arnold, 1996) treats science as if it is "not ... objective fact or authoritarian source, but as driven by various political and cultural agendas." Certainly, science and scientists are worthy objects of historical study, but, like Oliver-Smith (2015), I think they should not be "culturiz [ed]" to the point that "environment and nature ... disappear into a haystack of discourses." As those of us who have experienced the power of a hurricane know well, a cyclonic storm is a real physical phenomenon, not a mere rhetorical construction. And like Wisner et al. (2004), I think a balance can be struck between the environmental determinism of focusing too much on the triggering of natural hazards and the "equally deterministic approach" of some recent scholarship, in which disaster is "rooted in the political economy alone." This book will therefore try to find a happy medium between the underlying science of natural hazards and the numerous human cultural, economic, political, and social factors that transform these hazards into disasters.

What Is "Natural"?

Human beings face a bewildering array of hazards that threaten our lives and livelihoods. Some of these hazards are entirely man-made, such as the Great Molasses Flood of 1919, when a sticky wave of molasses erupting from a burst storage tank killed 21 inhabitants of Boston's North End. What is more, it is often argued that the most deadly and pervasive man-made disaster throughout human history has been warfare, which overall has probably claimed more lives and destroyed more property than all the "natural" disasters put together. Recurrent human warfare, a self-inflicted wound that runs like a red thread through the tapestry of human history, could be seen as the quintessential example of a *non*-natural disaster.

But is warfare really so unnatural? As Niall Ferguson pointed out in his 2021 book on disasters and human history, many historical wars were triggered in part by climate change and resultant agrarian crises, which compelled that society to choose between "starvation and relocation." Non-human animal species react to food scarcity in analogous ways. What is more, the roots of warfare lie at least in part in human biology. In an important but overlooked 2006 paper historian Walter Scheidel contended that warfare in the pre-modern world was driven largely by the (largely unconscious) sexual strategies of human males, who used war to increase their reproductive fitness in two different ways: by acquiring resources that could be used to attract wives, and by gaining opportunities for "marginal reproductive success" outside of marriage, either through war-time rape or the acquisition of female slaves. While any given war might also have political or ideological causes, Scheidel contends that "empires could not have been established in the absence of [male] behavioral traits that have evolved to promote inclusive fitness." What is more, while most of Scheidel's examples deal with warfare in the ancient world, he also notes that the young male soldiers who fought in European imperial expeditions probably had similar biological motives, as did the predominantly young and male migrants attracted to European colonial territories. The old adage that there would be no war in a world ruled by women, therefore, might contain a grain of truth.

Scheidel's argument about the "naturalness" of human military conflict receives strong support from the study of our closest genetic relatives, chimpanzees, who try to increase their reproductive fitness through behaviors that are closely analogous to human territorial warfare. According to military historian Doyne Dawson (1999), the males in chimp troops sometimes join together in bands to make raids into the territory of rival groups for the purpose of stalking and killing rival male chimpanzees. If these raids are successful, the victorious chimp troop will eventually annex the territory, and the females, of their vanquished rivals. There are obvious parallels between such "natural" chimp behavior and the common practice in early human warfare for the victors to seize the territory and the women of a vanquished enemy, but to kill off the defeated males.

However, such evidence about the biological roots of some human behavior runs counter to the dominant narrative of man/nature dualism we have inherited from the Judeo-Christian religious tradition. As Lynn White, Jr. pointed out in a seminal 1967 article, western science derived its view of the natural world from the medieval religious beliefs that human beings and the world were separate creations, that nature was given to man for human exploitation, and that man's proper place is thus above the natural world rather than within it. However, as we learn more about human genetics and evolution, this sharp distinction between man and the natural world becomes harder to sustain. So too does the corollary belief common in the social sciences, that human beings are born as blank slates with no intrinsic nature and that their behaviors are purely determined by cultural learning.

I am certainly not saying that culture doesn't matter. On the contrary: as we will see throughout this text, culture plays a crucial role in natural disasters, for example, by modulating human vulnerability to hazards and by determining how disasters are defined, interpreted, and remembered. However, as I will argue in Chapter 1, biology matters as well. Our shared genetic heritage means that human beings across the world tend to react to disasters in predictable, biologically conditioned ways, such as through the development of fast or slow life strategies and the activation of our pre-programmed "behavioral immune system" when faced with post-disaster disease cues. Chapter 1 will address the large and growing literature on this topic in the natural sciences, especially the emerging field of evolutionary psychology, that unfortunately has been almost completely neglected by the field of disaster studies.

Scope of the Book and Chapter Summaries

This text will explore the many themes outlined above by applying insights from the social sciences, along with material drawn from the natural sciences, to study specific case studies of natural disaster. Rather than trying to treat all possible natural hazards, the book will sacrifice some breadth for depth and focus on three hazards caused primarily by tectonic forces (earthquakes, volcanoes, and tsunamis), one hazard caused primarily by meteorological forces (cyclones), and one hazard, climate change, that can be triggered by meteorological factors, tectonic forces, and/or human agency. Although disease is an extremely common trigger for disaster, and many diseases are "natural" insofar as they are caused by biological agents, disease is such a complex phenomenon that it is best covered on its own in another book. Nonetheless, disease will sometimes become important in this text as a secondary side-effect of disaster triggered by a quake, cyclone, or other hazard. Much the same is true about famine, which can also be linked to natural causes, such as climate change. Specific famines will be covered in the text, but again this is a topic worthy of study in its own right and space does not permit full coverage of it here.

Since one aim of this book is to provide students with a clear explanation of the "natural" elements of a natural disaster, Chapter 1 will provide a broad overview of scientific processes that are important to understanding all disasters highlighted in this text. This chapter begins by examining the all-important process of convection and its relevance to both meteorology and earth science, since it explains both the existence of Hadley cells in the atmosphere and the movement of tectonic plates in the Earth's crust. Second, this chapter will turn to the human science of disasters, in particular, the common and recurring psychological impacts that disasters invoke in their victims. We will also examine several developmental and cognitive mechanisms produced by human evolution that are relevant to disaster studies. Case in point is the pan-human phenomenon of the behavioral immune system, which helps to explain why disasters often bring out the worst in us, triggering xenophobia, out-group exclusion, and even genocidal behavior against marginalized groups.

Once students have this background under their belts, they are ready to move on to Chapter 2, which examines natural disasters caused directly by earthquakes. This chapter will begin by explaining folk theories of earthquakes as well as scientific concepts specific to earthquakes, such as different types of tectonic faults and seismic waves. The chapter will then move on to multiple case studies of historical earthquakes, including thick descriptions of the 1775 Lisbon earthquake, the 1906 San Francisco quake, the 1923 Great Kanto earthquake and firestorm, and the relatively recent 2010 earthquake in Haiti, which may be the single deadliest single-event, non-disease disaster ever recorded in the Western Hemisphere.

Chapter 3 considers another disaster that is generally (though not exclusively) triggered by tectonic forces: the tsunami. Tsunamis waves are as old as recorded history – as we shall see, Homer's *Illiad* may make reference to a half-remembered tsunami of the distant past. Our main case studies in this chapter, however, are more recent: the Indian Ocean Tsunami of 2004 and the Tōhoku Tsunami of 2011. Both events are extremely well documented, allowing them to be studied in depth. At the same time, these two tsunamis struck societies with very different levels of both socioeconomic development and political stability, allowing for illuminating cross-disaster comparisons.

Tectonic forces are also responsible for volcanic disasters, the subject of Chapter 4. As this chapter shall reveal, however, volcanoes are a rather more complex phenomenon than quakes and tsunamis. Volcanoes take on a number of different forms and lead to multiple types of hazard, ranging from short-range threats (like magma flows to pyroclastic flows) to longer-term risks (like lahar landslides and climate change induced by volcanic ash and sulfates). This chapter is divided into two sections. The first examines two volcanoes, Krakatau (1883) and Mount Pinatubo (1991), each of which posed hazards primarily through immediate, short-term threats to people in the surrounding areas. The second section, in turn, examines volcanic eruptions that triggered historically significant long-term or global climate change, especially the well-studied 536 eruptions event involving one or more unknown volcanoes, the Laki eruption of 1783, and the Mount Tambora eruption of 1815, which was famously followed by a "year without a summer."

Climate change induced by volcanism serves as a good segue into Chapter 5, which examines the historical consequences of climate change triggered by radiative forcing, including anthropogenic changes to the global environment. As this chapter will illustrate, the climate of any given part of the world, and any given time in world history, is governed by a variety of variables. One such set of variables are the short-term climate oscillations tied to the atmosphere and ocean, such as the ENSO and the NAO, which are centered in the Southern Pacific and the North Atlantic respectively. Both processes exert a powerful influence on the weather over wide areas and both can lead to disasters in their extreme manifestations, mainly in the form of drought-induced famine. On a longer scale, changes in the amount of energy received by our planet from the sun have led to recurrent periods of warmer and colder climate, such as the Roman Optimum of 100 BCE to 200 CE, the Late Antique Ice Age centered in the sixth century CE, Europe's so-called Medieval Warm Period of roughly 900 to 1200, and the Little Ice Age of 1500–1700 or beyond. Today, however, the age of climate change driven by radiative forcing is more or less over, replaced by the modern age of anthropogenic climate change, also known as global warming. This chapter will end by considering the impact of rising global temperatures both on contemporary history and on the future of our species.

Finally, Chapter 6 will close out this rogue's gallery of natural disasters by discussing tropical cyclones, a phenomenon better known to American students by their regional moniker, hurricanes. Born in tropical seas, cyclones pose a recurrent threat to large parts of East Asia, South Asia, the Caribbean, the American East Coast, and even Australia. Unfortunately, cyclones pose multiple hazards to vulnerable human societies, ranging from freshwater flooding due to heavy rains, saltwater flooding through tsunami-like storm surge, and direct damage from their spiraling winds. This chapter will examine three main cyclone case studies, including the Great Caribbean Hurricanes of the eighteenth and early nineteenth centuries and the Bhola Cyclone of 1970 that served as the midwife for the birth of a new country, Bangladesh. This chapter will close with a deep discussion of Hurricane Katrina of 2005, an American national tragedy that epitomizes the degree to which "natural" disasters are rooted in human political, economic, and social vulnerabilities.

Further Reading

Bankoff, Greg. "The Historical Geography of Disaster: 'Vulnerability' and 'Local Knowledge' in Western Discourse." In Greg Bankoff, Georg Frerks, and Dorothea Hilhorst (Eds.), *Mapping Vulnerability: Disasters, Development and People*. London: Earthscan, 2007, pp. 25–36.

Bankoff, Greg. "Cultures of Disaster, Cultures of Coping: Hazard as a Frequent Life Experience in the Philippines." In Christian Pfister and Christof Mauch (Eds.), *Natural Disasters, Cultural Responses: Case Studies Towards a Global Environmental History.* New York: Rowman & Littlefield, 2009, pp. 265–284.

Bankoff, Greg. "In the Eye of the Storm: The Social Construction of the Forces of Nature and the Climatic and Seismic Construction of God in the Philippines." In Ben Wisner, J.C. Gaillard, and Ilan Kelman (Eds.), *Disaster Risk*, vol. 1. New York: Routledge, 2015, pp. 406–429.

Bankoff, Greg. "Remaking the World in Our Own Image: Vulnerability, Resilience, and Adaptation as Historical Discourses." In Greg Bankoff and Dorothea Hilhorst (Eds.), *Why Vulnerability Still Matters: The Politics of Disaster Risk Creation.* New York: Routledge, 2022, pp. 15–32.

Bankoff, Greg, Georg Frerks, and Dorothea Hilhorst (Eds.) *Mapping Vulnerability: Disasters, Development and People.* London: Earthscan, 2007.

Oliver-Smith, Anthony. "Haiti's 500-Year Earthquake." In Mark Schuller and Pablo Morales (Eds.), *Tectonic Shifts: Haiti Since the Earthquake.* Boulder, CO: Kumarian Press, 2012.

Pfister, Christian. "'The Monster Swallows You': Disaster Memory and Risk Culture in Western Europe, 1500–2000." *Rachel Carson Center Perspectives*, no. 1 (2011), pp. 1–23.

Scheidel, Walter. "Sex and Empire: A Darwinian Perspective." *Princeton/Stanford Working Papers in Classics*, 2006. Available at: web.stanford.edu/~scheidel/pub.htm

Wisner, Ben, J.C. Gaillard, and Ilan Kelman (Eds.) *Disaster Risk*, vol. 1. New York: Routledge, 2015.

Works Cited

Alexander, D. E. "Resilience and Disaster Risk Reduction: An Etymological Journey." *Natural Hazards and Earth System Sciences*, vol. 13 (2013), pp. 2707–2716.

Arnold, David. *The Problem of Nature: Environment, Culture, and European Expansion.* Malden, MA: Blackwell, 1996.

Bankoff, Greg. *Cultures of Disaster: Society and Natural Hazard in the Philippines.* New York: Routledge Curzon, 2003.

Bankoff, Greg. "Time Is of the Essence: Disasters, Vulnerability, and History." *International Journal of Mass Emergencies and Disasters*, vol. 22, no. 3 (2004), pp. 23–42.

Beck, Ulrich. *Risk Society: Towards a New Modernity.* New York: SAGE Publications, 1992.

Bergman, Jonathan. "Disaster: A Useful Category of Historical Analysis." *History Compass*, vol. 6, no. 3 (2008), pp. 934–946.

Braudel, Fernand. *The Mediterranean and the Mediterranean World in the Age of Philip II.* 2 vols. Oakland, CA: University of California Press, 1996.

Burnham, John C. "A Neglected Field: The History of Natural Disasters." *Perspectives on History*, April 1, 1988.

Burton, Ian, Robert W. Kates, and Gilbert F. White. *The Environment as Hazard.* New York: Oxford University Press, 1978.

Dawson, Doyne. "Evolutionary Theory and Group Selection: The Question of Warfare." *History and Theory*, vol. 38, no. 4 (1999), pp. 79–100.

Del Moral, Roger and Lawrence R. Walker. *Environmental Disasters, Natural Recovery, and Human Responses.* New York: Cambridge University Press, 2007.

Ferguson, Niall. *Doom: The Politics of Catastrophe.* New York: Penguin, 2021.

Hagan, Ryan. "Acts of God, Man, and System: Knowledge, Technology, and the Construction of Danger." In Jacob A. C. Remes and Andy Horowitz (Eds.), *Critical Disaster Studies.* Philadelphia, PA: University of Pennsylvania Press, 2021, pp. 32–50.

Hewitt, Kenneth. "The Idea of Calamity in a Technocratic Age." In Kenneth Hewitt (Ed.), *Interpretations of Calamity from the Viewpoint of Human Ecology.* New York: Routledge, 1983, pp. 3–32.

Hilhorst, Dorothea and Greg Bankoff. "Introduction: Why Vulnerability Still Matters." In Greg Bankoff and Dorothea Hilhorst (Eds.), *Why Vulnerability Still Matters: The Politics of Disaster Risk Creation.* New York: Routledge, 2022, pp. 1–12.

Horowitz, Andy. *Katrina: A History, 1915–2015.* Cambridge, MA: Harvard University Press, 2020.

Keys, David. *Catastrophe: An Investigation into the Origins of the Modern World.* New York: Random House, 1999.

Klein, Naomi. *The Shock Doctrine: The Rise of Disaster Capitalism.* London: Picador, 2008.

Knowles, Scott Gabriel. "Learning from Disaster? The History of Technology and the Future of Disaster." *Technology and Culture,* vol. 55, no. 4 (2014), pp. 773–784.

Knowles, Scott Gabriel. "Slow Disaster in the Anthropocene." *Daedalus,* vol. 149, no. 4 (2020), pp. 192–206.

Moore, Harry E. *And the Winds Blew.* Austin, TX: University of Texas Press, 1964.

O'Keefe, Phil, Ken Westgate, and Ben Wisner. "Taking the Naturalness out of Natural Disasters." *Nature,* vol. 260 (1976), pp. 566–567.

Oliver-Smith, Anthony. "Theorizing Disasters: Nature, Power, and Culture." In Ben Wisner, J.C. Gaillard, and Ilan Kelman (Eds.), *Disaster Risk,* vol. 1. New York: Routledge, 2015, pp. 267–287.

Oliver-Smith, Anthony. "Peru's Five-Hundred-Year Earthquake: Vulnerability in Historical Context." In Anthony Oliver-Smith and Susanna M. Hoffman (eds.), *The Angry Earth: Disaster in Anthropological Perspective.* New York: Routledge, 2020, pp. 83–97.

Perry, Ronald W. "What Is a Disaster?" In Ben Wisner, J.C. Gaillard, and Ilan Kelman (Eds.), *Disaster Risk,* vol. 1. New York: Routledge, 2015, pp. 11–33.

Pfister, Christian. "Learning from Nature-Induced Disasters: Theoretical Considerations and Case Studies in Western Europe." In Christian Pfister and Christof Mauch (Eds.), *Natural Disasters, Cultural Responses: Case Studies Towards a Global Environmental History.* New York: Rowman & Littlefield, 2009, pp. 17–40.

Pritchard, Sara B. "An Envirotechnical Disaster: Nature, Technology, and Politics at Fukushima." *Environmental History,* vol. 17, no. 2 (2012), pp. 219–243.

Quarantelli, E. L. "Disaster Studies." *International Journal of Mass Emergencies and Disasters,* vol. 5 (1987), pp. 285–310.

Remes, Jacob A. C. and Andy Horowitz (Eds.) *Critical Disaster Studies.* Philadelphia, PA: University of Pennsylvania Press, 2021.

Steinberg, Ted. *Acts of God: The Unnatural History of Natural Disaster in America.* Oxford: Oxford University Press, 2000.

Van Bavel, Bas, *et al.Disasters and History: The Vulnerability and Resilience of Past Societies.* New York: Cambridge University Press, 2020.

Warner, W. Lloyd. *The Social System of the Modern Factory.* New Haven, CT: Yale University Press, 1947.

White, Lynn Jr. "The Historical Roots of Our Ecologic Crisis." *Science,* vol. 155, no. 3767 (1967), pp. 1203–1207.

Wisner, Ben, Piers Blaikie, Terry Cannon, and Ian Davis. *At Risk: Natural Hazards, People's Vulnerability and Disasters.* 2nd edn. New York: Routledge, 2004.

1 The Science of Natural Disaster

In January of 1998, an unusual concentration of meteorological factors triggered an unprecedented catastrophe in the upper St. Lawrence River valley between the northern United States and Quebec. Moisture-rich air moving north from the Gulf of Mexico collided with an unusually cold high-pressure air mass in Labrador, and rather than moving eastwards as was typical of such systems, these warring fronts were locked in place by a strong and obstinate Bermuda high pressure system. The result was 80 hours of steady freezing rain over Eastern Ontario, parts of northern New York and New England, and most especially Southern Quebec. Ice accumulations of 10 centimeters or more toppled maple trees and made barns collapse, crippling the local economy, and in Montreal falling sheets of ice from downtown buildings rendered the roads impassable to cars and pedestrians alike. Worst of all, over 1,000 electrical transmission towers collapsed under the weight of the ice, leading to widespread and long-lasting power outages, especially in the worst-affected "triangle noir" or "dark triangle" east of Montreal. Over 4 million people were deprived of electricity at the height of the crisis, and at least 34 people died during these desperate times. While some of the storm's victims succumbed to injury or hypothermia, many more died of asphyxiation, killed by carbon monoxide poisoning as they huddled for warmth around faulty generators or household hearths with blocked chimney flues.

Power, and some degree of normality, were eventually restored thanks to the work of tens of thousands of soldiers, National Guardsmen, and utility workers. But for the women who were pregnant during the Quebec Ice Storm, the damage was already done. In the last two decades, a series of studies have linked maternal prenatal stress caused by the Quebec Ice Storm to a whole suite of developmental problems among ice storm babies, including smaller overall size and reduced head circumference, lower IQ and language development, childhood obesity and type-1 diabetes, earlier female puberty, and even autism spectrum behavior, especially among males. What is more, these problems are not just growing pains, but are expected to persist. Genetic studies have

DOI: 10.4324/9781003436805-2

detected numerous changes to the genes of ice storm children through DNA methylation, a process by which certain genes are turned on or off *in utero* by methyl molecule tags, so many of these epigenetic alterations might last a lifetime.

In the end, the main lesson we can draw from the tragedy of the Quebec Ice Storm is the important role played by both natural science and human biology in the causes and consequences of disaster. After all, the freezing rains of January 1998 not only ravaged Quebec's economy and infrastructure, they reached deep into the human genome itself.

The Science of Natural Disaster

As the later case study chapters of this book will demonstrate, fully understanding how natural disasters unfold requires us to have at least some familiarity with a number of scientific concepts. Many of these concepts are specific to particular types of disaster, so it makes more sense to hold off on discussing them until we turn to the appropriate chapter. What this chapter will cover is some general scientific principles that are important to multiple disasters. In terms of natural science, the most important such principle is convection, a process that lies at the heart of cyclone formation, the movement of tectonic plates, the formation of volcanoes, and the rains of the tropical monsoon. As for human science, the scientific principles most relevant to disaster studies revolve around human psychology and development, especially the species-typical ways in which we adapt to challenging environments and react to disease cues. These psychological factors help to explain why humans react to natural disasters in predictable ways regardless of time period and cultural differences.

Natural Science and Disaster: Convection

In some ways, convection is the 2x4 Lego brick of the natural sciences: a simple building block that lies at the base of much more complex structures. We all have personal experience with convection, which occurs whenever fluids and gases are unequally heated. The movement of the wax blobs in a lava lamp, which rise up once they have absorbed enough heat from the light bulb at the bottom of the lamp, is the result of convection. Heated air wants to rise as well, since it is less dense and thus more buoyant than cold air, a fact much appreciated by fans of hot-air ballooning.

But as the expression goes, what goes up must come down. Whether the rising material is atmospheric air or magma (molten rock) in the Earth's crust, as it ascends and moves away from the heat source, it loses heat to surrounding matter, slowing its upward motion. Once this rising material reaches temperature equilibrium with the surrounding material, it begins to sink back downwards, displaced by warmer and less dense material upwelling from below. In the process, both air and magma release the gases they are carrying upwards with them: water vapor in the case of air and a mix of carbon dioxide, sulfur dioxide, water vapor, and other gases in the case of magma. As we will see in later chapters, these released gaseous materials are crucial to the science formation of cyclones and volcanoes respectively.

The sum result of these temperature-driven movements is the creation of convection cells, large-scale structures both in the atmosphere and the Earth's interior that in cross-section resemble turning wheels. Hurricanes are basically large and mobile convection cells, powered by heat from the ocean below and spun into circular horizontal motion by the Coriolis effect, which we will discuss below. Even larger convection cells power the phenomenon of plate tectonics, which accounts for the movement of the plates that form the Earth's surface, and form Hadley cells in the atmosphere that govern much of the world's weather.

Mantle Convection and the Theory of Plate Tectonics

The idea that the Earth's solid surface rests upon a fluid interior is as old as Benjamin Franklin, who speculated that our planet's crust was a solid outer shell atop a liquid and that earthquakes were the product of "violent movements of the fluid on which it rested." Early natural philosophers also speculated that the apparently stable continents of the world might have split and fused over geological time. This would explain, for example, why the coasts of Africa and South America would lock together like pieces of a puzzle if South America were somehow moved across the Atlantic and why the fossils found in Brazil so closely resembled fossils from Africa's Bight of Biafra. However, early scientists knew of no force that could have pulled South America from Africa's gasp. It was not until the 1950s and 1960s that scholarly agreement coalesced around a possible mechanism that explained continental drift: the theory of plate tectonics.

According to plate tectonics theory, the fundamental force re-shaping the solid surface of the Earth is the fluid movement of magma currents in the upper mantle. These mantle currents are, of course, the result of convection, the rise and fall of unequally heated rocky materials in the semimolten mantle beneath the lithosphere, or crust. Convection currents in the mantle slowly shift the continental and oceanic plates above, grinding them against, over, or under each other if the currents in the underlying mantle are traveling at different speeds or in different directions. As we will see in later chapters, these plate-on-plate interactions are the main factor responsible for earthquakes and tsunamis.

Upwelling magma re-shapes the face of the Earth in another way as well: if it reaches the surface, it will form a volcano. In technical terms, a volcano is any vent that allows magma or gases from the mantle to reach the surface. This happens most commonly at "normal faults" where two plates are being pushed apart from each other, creating an opening for rising magma to emerge and solidify. Volcanoes of this sort are most typically found deep in the ocean, where they form long mid-oceanic ridges made up of recently emerged volcanic rock. Such ridges can be thought of as massive 3-D printers, constantly building new sections of ocean floor out of basalt and other dense igneous rocks emerging from the mantle. Since these new crust materials are being formed in the ocean's centers, the older ocean crust is constantly being pushed outwards like a conveyor belt toward the lighter rock that forms the continents. At the point of collision between oceanic and continental plates, the former generally dives under the latter, a process called subduction. As we will see in the later chapters, subduction faults are responsible for our world's strongest earthquakes, the largest tsunamis, and the most dangerous type of volcano, the stratovolcano.

It is important to note that the "movement" of these plates is far from constant. For most of the time, tectonic plates are held together by friction, like the two interlocked pieces of a Velcro clasp. As a result, the accumulating kinetic energy from the motion of the mantle currents is stored within the plates as tension or potential energy. Movement of the plates only occurs when this tension builds up sufficiently to overcome the friction between the plates. At this point the Velcro rips, allowing for sudden and sometimes dramatic movement along the plate boundary. Humans experience this energy as an earthquake, if the energy that is released displaces the land, a tsunami if it displaces the ocean, or both. The total amount of energy that can build up along the plates depends on how they are interacting: it is lowest in "normal" faults where plates are spreading apart, higher in "transform" faults where the plates are moving alongside each other, and highest when two plates are crashing together. This latter category includes both subduction faults, when an ocean plate descends under a continental plate, and "thrust" faults where two continental plates are driven into each other. This latter process causes both plates to fold and buckle, like the front ends of two cars involved in a head-on collision, creating tall mountain ranges like the Alps, the Pyrenees, and the Himalayas. Plate collisions of this latter type produce frequent earthquakes but few volcanoes, since they leave no avenue for magma to reach the surface.

Plate tectonics is therefore a sculptor, constantly shaping and re-shaping the planet's surface according to the whims of the convectional forces of the mantle below. If this analogy is accepted, it helps to explain why the disaster risks caused by tectonic forces are unevenly distributed across the globe. The chisel and hammer of plate tectonics are not equally active everywhere but rather are focused on the boundaries between tectonic plates, particularly along subduction and thrust faults and in areas where the underling mantle movements are strongest. Such regions include the Mediterranean, the Caribbean, the northern rim of the Indian

Subcontinent, and, above all, the Pacific "Ring of Fire," which encompasses Japan, the Philippines, Indonesia, and the western coast of the Americas. However, the work of the sculptor sometimes spills beyond those areas and impacts wider regions, either due to the long-term transfer of kinetic energy through a tsunami wave or through regional or global climate change brought on by volcanic ash and sulfates.

Atmospheric Convection, Hadley Cells, and the Coriolis Effect

At the same time that convection in the mantle is continuously re-shaping the Earth's crust, convection in the atmosphere is constantly remaking the Earth's weather, though on considerably shorter time scales (Figure 1.1). Atmospheric convection is driven by solar energy, which is strongest near the equator as this is where the sun's rays hit the Earth at closest to a 90° angle. When this strong, direct sunlight strikes the equatorial lands and oceans, it is absorbed and transformed into heat energy, fueling a rising column of air heavily laden with moisture. As this air column soars skyward and begins to cool in the upper atmosphere, the moisture precipitates back out, generating the towering clouds and downpouring rains for which the equatorial region is famous. The superheated spot below the sun where columns of warm, moist air constantly rise into the upper

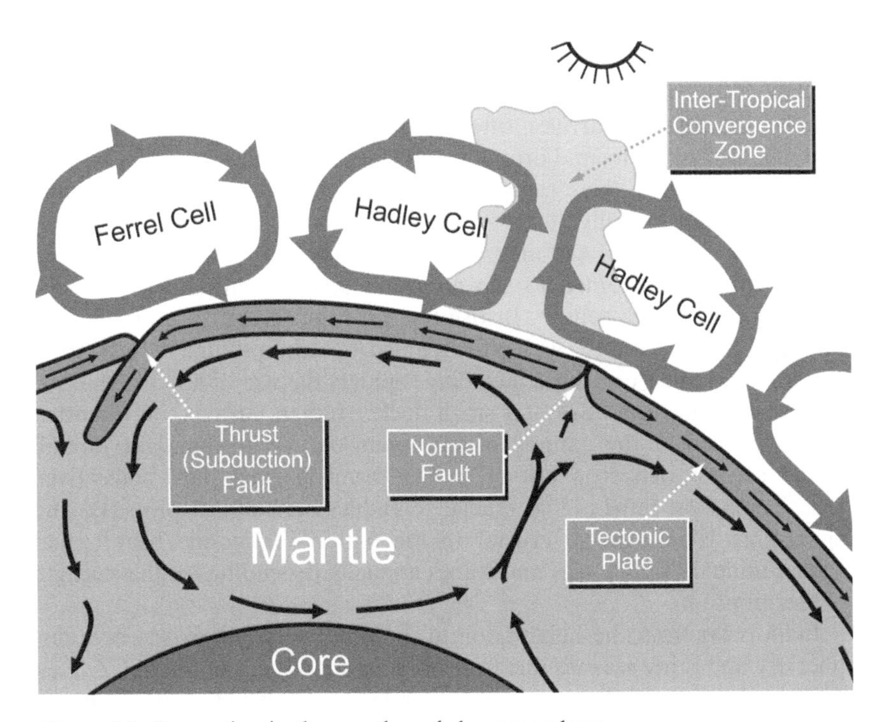

Figure 1.1 Convection in the mantle and the atmosphere

atmosphere is called the Inter-Tropical Convergence Zone, or ITCZ, and is often visible on satellite maps as a belt of dense rainclouds encircling the midriff of the planet.

The ITCZ is also the meeting point of two huge atmospheric convection cells, one on each side of the ITCZ, which between them cover a substantial proportion of the surface of the Earth and govern much of its weather. These convection cells, called Hadley cells after their discoverer, are far from static. As the Earth passes through its 12-month orbit around the sun, the portion of the Earth that is at a 90° angle to the sun constantly shifts. In mid-summer, when the Earth's Northern Hemisphere is tilted toward the sun, the northernmost point at which the Earth directly under the sun lies at about 23.4°, a latitude known by geographers as the Tropic of Cancer. During the Northern Hemisphere's mid-winter, it is the Southern Hemisphere's turn to lie directly below the sun, which shines down upon the Tropic of Capricorn at 23.4° south latitude. Over the course of the rest of the year, the ITCZ migrates between these two points, crossing the equator twice in the process, in mid-fall on its way to the south and in mid-winter as it moves back to the north.

When the ITCZ arrives in an area over the course of its seasonal migrations, the impact can be dramatic. Before the monsoon arrives in late May, for example, India's fields and farms have been baked a sterile brown and melting asphalt squelches beneath truck tires in the 40° Celsius heat. Then, sometime in early June, clouds begin to roll in from the south, thickening over time into dense gray mats. The humidity level creeps toward 100%, making the already-hot temperatures almost unbearable. Finally, the first rain arrives, often in the form of a torrential downpour that can last for hours and drop the ambient temperature by a full 10°C or more. Once started, these rains typically last for three to four months and in sum account for fully 70% of India's total annual rainfall.

With the coming of the monsoon, India's landscape metamorphosizes. Europeans unaccustomed to the monsoon season were shocked by the changes; seventeenth-century British traveler John Ovington wrote that the "time of the Mussouns" was accompanied by "the prodigious growth of vermin" and other creatures, including "spiders the size of a man's thumb" and toads nearly the "size of a small duck." Indian farmers, on the other hand, rejoiced at the coming of the monsoon, when abundant rainfall would coax a rich crop out of India's formerly brick-hard soils. Even today, monsoon rains are so crucial to India's agricultural property, and agriculture is in turn so crucial to the nation's economic health, that India's annual GDP waxes and wanes annually depending on the strength of the monsoon.

India is far from the only region in the subtropics that experiences distinct dry and rainy seasons due to the annual movement of the ITCZ. East Asia is also impacted by monsoon thunderstorms, which reach southern China and northern Vietnam by the end of May and then Japan and northern China by mid-July. The African "Sahel" region, a savanna zone south of the Sahara Desert, receives most of its annual moisture from the monsoon during the summer season, as does the Horn of Africa region to the east. When the ITCZ moves south of the equator during the winter, it leaves these areas high and dry, but carries monsoon moisture as far south as northern Australia and Madagascar.

Nor is the influence of the Hadley cells limited to the tropics. In the subtropical end of the Hadley cell, air descending from high altitudes contributes to prevailing high pressure and low humidity conditions. As a result, the subtropical latitudes are almost synonymous with deserts, including the Sahara, Arabian, Thar, and Mohave deserts in the Northern Hemisphere and the Kalahari and Australian Deserts south of the equator. Ironically, the Hadley circulation can also pump life into the same deserts; the Nile River, for example, is fed almost completely by monsoon rain falling in the Ethiopian highlands 2000 kilometers to the south. In addition, the Hadley cells influence weather even further to the north and south via their interaction with Ferrel cells, atmospheric circulations that dominate the mid-latitudes of the Northern and Southern Hemispheres. Since the ground-level air in Ferrel cells flows northwards from the subtropics, Ferrel cells serve to pump heat and moisture from the subtropical zones into higher latitudes north and south of the equator.

In addition to redistributing heat and moisture worldwide, these atmospheric circulation cells also determine the direction of the prevailing winds, though this process is complicated by the rotation of the Earth. Surface air in the Northern Hemisphere, for example, is constantly being pulled southwards toward the relative vacuum formed by the rising column of air at the ITCZ. From the standpoint of an observer on the ground, however, that same air seems to be traveling, not straight south, but curving westward. This is the result of the Coriolis effect, a by-product of the differential between the long distances traveled by the surface of the spinning Earth at the equator, which is the largest of the latitude lines, vs. the shorter distances traveled by the Earth at higher latitudes. As a result, any weather phenomenon large enough to occupy a substantial portion of the Earth's surface is deflected in a clockwise direction in the Northern Hemisphere and counterclockwise in the south.

The Coriolis effect is important to the study of disasters in several ways. The spin that the Coriolis effect imparts to developing surface-level low-pressure cells is crucial to cyclone formation. It also explains why cyclones rotate clockwise in the Northern Hemisphere and counterclockwise to the south as well as the tendency of cyclones to "recurve" to the right in the Northern Hemisphere. What is more, the increase of the rate of deflection as you move toward the poles helps to explain why cyclones typically do not form near the equator: at these latitudes, the influence of the Coriolis effect is too weak to give a rotational spin to a convection cell. What is more, the Coriolis effect also explains why the prevailing winds in most parts of the world flow consistently in one direction. The higher latitudes are dominated by "westerlies," which (confusingly) track consistently to the east, while westward-flowing "trade winds" dominate the equatorial regions. These trade winds are also important because they contribute to the ENSO (El Niño/Southern Oscillation) cycle, a major engine of short-term climate change.

If plate tectonics is imagined as a sculptor, therefore, a Hadley cell should be likened to a painter, one who uses heat and humidity to color the terrestrial world with various shades of blue, green, white, and brown. But Hadley cells are fickle painters. The path and the strength of the ITCZ shift from year to year as well as from century to century, as solar forcing, major volcanic eruptions, and other factors shift its path. Over geological time, changing climate has dabbled lakes into the Sahara, painted rivers across Arabia, and slathered several kilometers of glacial ice over parts of Europe and North America. What is more, as we will see in Chapter 5, recent and more moderate climatic shifts over time have played a crucial but unappreciated role in shaping human history.

Natural Disaster and Human Biology

As the section above makes clear, natural science processes play an important role in a variety of different natural disasters. Given the state of the literature on disaster studies, this is not a particularly controversial statement, though it might invite some pushback from critical disaster study scholars. In the remainder of this chapter, however, I want to push the point further and argue that fully understanding the origins and outcomes of natural disasters requires us to consider not only natural science, but also human science, particularly psychology.

Turning first to some low-hanging fruit, it is widely accepted that natural disasters can lead to pervasive mental health disorders. As will come as no surprise to most of who lived through the COVID-19 pandemic, for example, the prevalence of depression and anxiety worldwide spiked by a "massive" 25% during the pandemic's first year, at least according to a scientific brief by the World Health Organization (2022). Similarly, a literature survey recently published in the *Chonnam Medical Journal* by Ju-Yeon Lee et al. (2022) found that between 3% and 37% of the survivors of both man-made and natural disasters suffer from significant levels of post-traumatic stress disorder, or PTSD. From the standpoint of the wider society, these mental health problems are both unfortunate and unproductive, as they inhibit people from assisting in post-disaster recovery and returning to normal life when the disaster ends.

While it may seem paradoxical that human biology complicates our ability to recover as a society from disasters, we should keep in mind that human behaviors evolved to maximize individual fitness, not to promote the interests of wider society. Indeed, one school of thought is that mental "disorders" are not really disorders at all, but rather evolved mechanisms for human self-preservation. According to evolutionary psychologist Randoph Neese (2016), for example, anxiety can be useful during crisis situations as it increases vigilance and speeds up our reaction times. Anxiety also triggers a strong "fight or flight" response that might encourage us to escape from hazardous situations. What is more, Neese argues that PTSD is certainly unpleasant for the sufferer but could also prove adaptive, since it represents "one-time learning of the strongest sort" that might remind and thus protect us from the recurrence of past dangers. As for depression, its adaptive function is less obvious, but it may serve to predispose us to self-protective inaction in situations where action could be dangerous, such in the presence of a hostile and physically superior rival or in the face of an impossible situation. As Nesse reminds us, our genes are ultimately interested, not in our short-term happiness, but in our long-term survival and reproduction.

Disasters and the Behavioral Immune System

Evolution may also offer an explanation for an often-observed historical phenomenon: the rise of xenophobia and the scapegoating of outsiders in the wake of a disaster. Consider, for example, four major disasters: (1) the bubonic plague epidemic in late medieval Europe; (2) the 1899–1901 Boxer Rebellion in China; (3) the Japanese Great Kantō earthquake and firestorm of 1923; and (4) the Haitian earthquake of 2010. On the surface, these crises seem to have little in common, as they are disconnected in space and time and were caused by quite different mechanisms. However, they have a surprising commonality: during all four events, outsiders were accused of poisoning community wells and were targeted for brutal murder. In the case of the bubonic plague, the fury of the mob descended upon Europe's Jewish communities. During the Boxer Rebellion, charges of well poisoning were leveled against Christian missionaries and their Chinese converts, both of whom were subjected to gruesome executions. In the Great Kantō quake, the well-poisoning scapegoats were Korean immigrants living in Japan, thousands of whom were slaughtered by Japanese vigilante groups in the quake's aftermath. Finally, in the months after the Haitian earthquake, as many as 45 Vodou priests in Haiti were lynched due to wild rumors that they had deliberately triggered a post-quake cholera epidemic by poisoning wells with magic powders.

Of course, each of these events has its own deep history, and none of these killing sprees can be reduced to a single cause. Nonetheless, I suspect that in all the cases above, the massacres were rooted to at least some degree in a psychological mechanism that all human beings share: the behavioral immune system (BIS). According to Mark Schaller (2011; 2016), who first introduced this now well-studied concept, human beings, like other animals, use two different mechanisms to protect themselves from infection. One is the better-known biological immune system response, which comes on-line when the body detects the intruding presence of pathogens or parasites and which can now be strengthened in many cases through the use of vaccines. However, this internal immune system is biologically costly and did not guarantee the survival of the organism, especially in ancestral environments.

As a result, many animals have also developed a *behavioral* immune system, which employs deeply ingrained emotions, especially disgust, to encourage protective behaviors and attitudes, such as aversion to diseased locations or sick populations as well as general hostility to out-group members. The behavioral immune system also sensitizes us to purity and possible contamination, especially in our food and water, which may explain why rumors of well poisoning are so common after major disasters. Disgust and aversion can also be extended to human beings, such as out-group members and social deviants. These behaviors may have been adaptive in ancestral environments, Schaller claims, in part because both foreigners and social non-conformists were more likely than in-group members to host novel pathogens against which local populations lacked any immunological defense.

However, such behaviors have their own costs, both in terms of lost energy and lost opportunity. Schaller therefore hypothesized that the behavioral immune system does not operate constantly, but rather is being continuously modulated by the magnitude of perceived disease risks within the organism's immediate environment. Sensitivity to disease cues also increases or decreases based on an organism's vulnerability to disease. Since a woman's biological immune system is partially suppressed during the first trimester of pregnancy, lest her own immune system target the fetus within her, the BIS tends to be easily activated during the first trimester of pregnancy. Newly pregnant women are more likely to feel disgust (this is the likely origin of morning sickness) and are more easily induced to in-group preference and out-group animus than non-pregnant women in laboratory settings. Research also suggests that people who personally feel susceptible to disease tend to overperceive disease cues and thus are more easily induced to xenophobic behavior. What is more, although there is only limited research at present to back it up, it is theorized that hunger and malnutrition may also modulate the behavioral immune system, though in complex ways. Valerie Curtis et al. (2011) predict that malnutrition would "downregulate disgust for food," but since malnutrition compromises the biological immune system, the BIS would become more vigilant in response to "other potential sources of pathogens," such as contamination and disease cues. Since natural disasters often trigger famines, it is reasonable to believe that starving survivors of such events would be especially sensitive to disease cues and thus predisposed to exclusionary and xenophobic behavior.

But how are disease cues actually perceived? Most pathogens after all are far too small to be detected by the naked eye. As a result, Schaller argues that our brain assesses these risks through proxy cues of disease: the reek of decomposition, the stink of fecal matter, and the appearance of physical symptoms in our fellow humans, such as coughing, vomiting, skin discoloration, and other bodily abnormalities. Aversion to such disease cues is not learned, but intrinsic (Figure 1.2). Curtis et al. (2011) argue that human populations worldwide recognize and react to the characteristic facial expressions associated with disgust without the need for a common culture or language. What is more, cross-cultural studies establish that all human populations react predictably with disgust when exposed to "bodily wastes, body contents, sick, deformed, dead or unhygienic people, [and] dirty environments."

Disease cues of the type that Curtis is describing are of course ubiquitous in the aftermath of any deadly and destructive natural disaster, which may help to explain why xenophobic outbursts are commonly seen in disaster's wake, as epitomized in the four examples outlined above. Pointing out this general tendency, however, should not be taken for a nod toward biological determinism. As recent work by Samuel Cohn (2018) has illustrated, at least in the case of disease disasters, scapegoating and massacres are in no way inevitable in the aftermath of catastrophe, and xenophobic impulses can be mitigated or even negated by political action or cultural beliefs. Case in point is the 1755 Lisbon earthquake, where simmering anti-foreigner sentiment never boiled over into murder or massacre on a large scale despite abundant disease cues due in part to the quick and decisive action of the Pombal regime. We'll consider this example in more depth in Chapter 2.

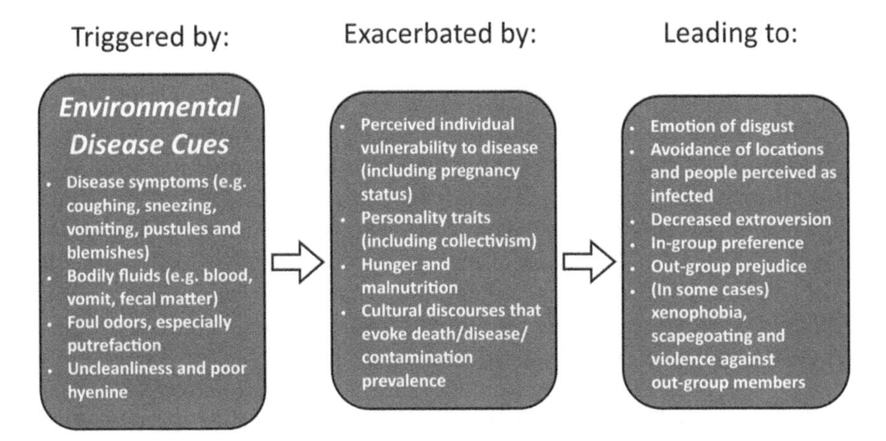

Figure 1.2 The behavioral immune system

Disaster and Life History Theory

Another emerging concept from the field of sociocultural psychology that has relevance to the study of natural disasters is life history theory, which postulates that humans and other organisms make strategic decisions about present and future resource allocations based on predicted environmental conditions. Organisms pursuing a "slow" life history strategy tend to focus on future pay-offs – they mature later, pursue stable relationships, invest more resources into fewer offspring, and take fewer risks. The opposite is true for "fast" life strategy organisms, which focus on immediate rather than future gains, are more tolerant of risks, become sexually mature earlier, have more sexual partners but less attachment to those partners, and invest less into more offspring. Different species show clear tendencies toward one or the other strategy – bacteria and most insects are "fast" life strategy organisms, while whales, elephants, some birds, higher primates, and humans lie at the "slow" end of the spectrum. In addition, there can be considerable variation within a species itself, with some individuals adopting faster strategies than others.

It is important to note that neither strategy is intrinsically better than the other. Rather, the success or failure of a strategy depends on the environment. Slow life strategies are more advantageous if environmental conditions are constant and predictable, where it pays for an organism to fully mature before breeding, to specialize in the exploitation of specific niches, and rigorously prepare for the future. However, in fickle and changing environments, a faster life strategy that exploits the various opportunities of the here and now might prove to be more adaptive. There is no point preparing for a future that may never come, or investing heavily in offspring that are unlikely to survive.

Among humans, at least, the evidence suggests that the choice of life strategy is not biologically determined. Rather, human beings show significant plasticity in this regard, able to pursue either strategy as conditions warrant. Research suggests that humans begin to gravitate toward a particular strategy before the age of 5 based on environmental cues, which we use subconsciously to predict the likely environment we will face as adults. Harsh parenting, childhood exposure to violence, childhood observation of death and disability, and erratic environmental conditions are all early-life cues that predispose individuals toward faster life strategies. So is low childhood socioeconomic status. According to Marco Del Guidice et al. (2016), children from poor families are more likely to pursue fast life history strategies as adults; they mature earlier, choose to have children at younger age at the cost of their own education and career development, take more financial risks, and save less for the future.

So what does life history strategy have to do with natural disasters? Perhaps quite a lot. Although the choice of a fast or slow life strategy in humans seems to be primarily linked to an unstable home life, the research suggests that erratic environmental conditions and exposure to death and misery play a role as well, so exposure to early childhood natural disasters could predispose individuals or even entire societies in the direction of fast life histories. Indeed, according to social psychologist Daniel J. Kruger and his colleagues, life history strategy might help to explain the apparent "preparedness paradox" – the well-studied fact that people living in unstable environments tend to report a high level of concern about possible disasters but do not translate that concern into a proportional level of disaster preparedness (Kruger et al., 2019). If people living in disaster-prone areas are more likely to adopt faster life strategies due to childhood exposure to uncertain environmental conditions, Kruger et al. argue, then this "paradox" is not really a paradox at all. People who have adopted faster life history strategies may feel like "potential crises may be too numerous and diverse" to be countered, making it "impossible to prepare for all contingencies." Fast life strategies, after all, predispose people to become generalists, not specialists, and to avoid investing too much into any specific domain. In an environment rife with uncertain natural hazards, Kruger et al. speculate, it might make more sense for fast-strategy individuals and populations to maintain an overall state of alertness against all possible threats rather than investing heavily against a single possible source of future disaster.

This work by Kruger et al. has clear implications for the study of "cultures of disaster," as described by Bankoff (2003) and others. Many of the traits that Bankoff ascribes to the Filipino culture of disaster, such as their diverse sources of income, their patterns of agricultural land fragmentation, their fatalism, and their normalization of risk, are comprehensible within the framework of fast life strategies. What is more, the idea of the "preparedness paradox" is strongly applicable to the Philippines, where anxiety about disaster is not matched with a commensurate level of disaster preparation. Bankoff quotes a columnist for the *Manila Chronicle* on exactly this point: "in a country beset by calamities most of the time, any planning is hard to do, and (the argument goes) it might as well be forsaken. The culture becomes a reactive one, prodded along by events and not mastering its future." Life history theory, therefore, gives some scientific grounding to the idea of "cultures of disaster" and might provide historians with tools for better understanding the mechanisms by which uncertain and hazardous environments influence the political, economic, and cultural systems of the societies they host.

Tightness, Collectivism, and Disaster

Historians who study cultures of disaster could also benefit from some other recent work in the field of socioecological psychology, including the concepts of cultural tightness/looseness and cultural collectivism/individualism. "Tightness" and "looseness," as Michele Gelfand and her colleagues pointed out in an influential 2011 article in *Science*, is a measure of a society's willingness to tolerate deviance from established social and cultural norms (Gelfand et al., 2011). Cultures across the world show considerable variation along this axis. Some countries like India, Pakistan, and Malaysia are extremely "tight," they expect strong adherence to social norms, discourage signs of public dissent, tolerate more autocratic systems of government, and invest more into police and judicial institutions. People in tight states also show more religiosity, both in terms of professing belief in God and attending regular religious services. In contrast, "loose" countries such as the Ukraine, Israel, and the Netherlands have less restrictive social norms, weaker religious and policing institutions, and afford much more autonomy to the individual. What is more, tightness-looseness scores can vary considerably even within a single region. A study by Jesse Harrington and Gelfand (2014) found that U.S. states exhibited considerable variation on tightness-looseness scores, which ranged from a very tight 78.86 in Mississippi to 27.37 in uber-loose California.

So why do different regions and sub-regions vary so much in their willingness to tolerate non-conformity and dissent? At this point in the research, it is hard to say what the ultimate cause or causes of the difference might be, but researchers have identified a number of factors that correlate with cultural tightness, including high population density, historic frequency of conflict, and higher crime and incarceration rates, though it is not clear if this latter factor is a cause of tightness or an artifact of more stringent controls over the population. Of more relevance to the present study is the finding that tightness is strongly correlated with ecological threats to a society, including the frequency of disease and ecological disasters. In their American study, Gelfand and Harrington found that tighter states had higher death rates from heat, storms and floods, and tornados than loose states, though this correlation did not hold true for deaths due to earthquakes and cold temperatures. Tightness is also "positively related to all indicators of disease prevalence reported to the US Census Bureau." The authors of this study do not go so far as to claim that this correlation amounts to causation, but such a conclusion would be in keeping with their theoretical perspective. After all, as Gelfand et al. noted in 2011, the main advantage of tightness is that it provides social elites with the strong norms and punishments they need to coordinate the population against threats to the society, including both "natural disasters" and the "spread of disease." If this is true, it highlights another mechanism by which ecological adversity might contribute to the formation of a culture of disaster.

Natural disasters are implicated in another cultural continuum as well, the better-known distinction between collectivist and individual societies. As described by Dutch psychologist Geert Hofstede in the 1980s, collectivistic societies prioritize the needs of the group over the individual, favor personality traits and attributes like harmony and social cohesion, and show greater preference for in-group members. Individualist societies, in contrast, value personal independence, self-achievement and self-sufficiency, prioritize individual rights over the rights of the group, and exhibit less in-group favoritism. As with the tightness/looseness continuum, considerable variation along this gradient exists both between countries and within them. Worldwide, Asian nations tend toward collectivism, Northern European and North American societies toward individualism. There can also be considerable variation within these countries; the mountain states of the United States epitomize individualism, while most of the old Confederate states strongly lean toward collectivism, and in China greater individualism prevails in the north and collectivism in the south.

So where does collectivism come from? One fruitful line of research has linked collectivistic social behaviors in modern societies with the historical disease frequency. According to Corey Fincher et al. (2008), countries occupying historical environments with high levels of disease and parasite prevalence are far more likely to exhibit modern-day collectivistic attitudes, presumably because collectivist behaviors such as in-group preference may have served an "antipathogen defense function" in the past. Collectivism is hypothesized to be protective against disease in part by discouraging contact with out-group members, who historically have been more likely to host exotic pathogens. In addition, collectivist societies exhibit greater social conformity, which may reinforce prophylactic social norms. William McNeill (1977), for example, has noted that the Tamils of southern India had a strong traditional taboo against storing unused drinking water within the home, a protective cultural practice which prevented malaria and dengue-carrying mosquitos from breeding within human habitations. Similarly, traditional Manchurian societies developed taboo laws about hunting fur-bearing marmots that minimized exposure to sick animals, thus reducing the possibility of cross-species infection by the *Y. pestis* bacillus, a pathogen endemic to marmots that causes deadly bubonic plague in humans. Collectivism reinforces the protective effect of such cultural practices by increasing social conformity as well as encouraging greater sanctions against norm-violating individuals.

In a similar vein, psychologists Shigehiro Oishi and Asuka Komiya (2017) have argued that collectivism may be an adaptive response to historical natural disaster prevalence. "Since their earliest time on Earth," these authors contend,

> humans have had to deal with natural disasters such as earthquakes, floods and volcanic eruptions. Disasters are part of human life and history. Humans have created certain norms and values to deal with the vicissitudes of life. [As a result] nations with a higher natural disaster risk tend to be more collectivist[ic] perhaps because group cohesion is an adaptive response to such a disaster.

That being said, Oishi and Komiya concede that their statistical analysis indicates that other factors, especially modern GDP per capita and historical disease prevalence, also seem to play a significant role in predisposing societies toward collectivism.

This chapter has tried to explain some of the various natural science concepts that are generally relevant to the origins and outcomes of the natural disaster types discussed in this text. As we have seen, scientific principles help to explain not only the shape of the continents, the origins of earthquakes, the rise of volcanic mountains, the patchwork layout of forest and desert, and the twirl of cyclones, but also some evolved mechanisms underlying out-group aversion, pro-social conformity, and childhood development. But, to borrow an analogy made by historian David T. Courtwright (1996), trying to understand human history solely through biological evolution and other underlying scientific processes is like trying to tell time using only the hour hand of a clock. Fully explaining any natural disaster requires us to understand not only general scientific principles, but also specific human contexts, both in terms of intermediate-term socioeconomic, political, and cultural trends (which could be likened to the minute hand of a clock) and shorter-term *histoire événementielle* (the seconds hand). While this chapter looked mainly at the hour hand, the chapters that follow will put the minute and second hands at center stage, examining in various case studies how specific human contexts have influenced the origins and outcomes of earthquake, tsunami, volcanic, climatic, and cyclonic disasters.

Further Reading

Bjorklund, David F., Carlos Hernández Blasi, and Bruce J. Ellis. "Evolutionary Developmental Psychology." In David M. Buss (Ed.), *The Handbook of Evolutionary Psychology*, vol. 2. Hoboken, NJ: Wiley, 2016, pp. 904–924.

Cao-Lei, Lei *et al.* "DNA Methylation Signatures Triggered by Prenatal Maternal Stress Exposure to a Natural Disaster: Project Ice Storm." *PLoS ONE*, vol. 9, no. 9 (2014). doi:10.1371/journal.pone.0107653.

Dancause, Kelsey N. *et al.* "Disaster-Related Prenatal Maternal Stress Influences Birth Outcomes: Project Ice Storm." *Early Human Development*, vol. 87 (2011), pp. 813–820.

Duchesne, A. *et al.* "Childhood Body Mass Index at 5.5 Years Mediates the Effect of Prenatal Maternal Stress on Daughter's Age at Menarche: Project Ice Storm." *Journal of Developmental Origins of Health and Disease*, vol. 8, no. 2 (2017), pp. 168–177.

Faulkner, Jason, Mark Schaller, Justin H. Part, and Lesley A. Duncan. "Evolved Disease-Avoidance Mechanisms and Contemporary Xenophobic Attitudes." *Group Processes and Intergroup Relations*, vol. 7, no. 4 (2004), pp. 333–353.

Figuerdo, Aurelio José, Michael A. Woodley of Menie, and W. Jake Jacobs. "A General Factor of Personality: A Hierarchical Life History Model." In David

M. Buss (Ed.), *The Handbook of Evolutionary Psychology*, vol. 2. Hoboken, NJ: Wiley, 2016, pp. 943–967.

Frerichs, Ralph R. *Deadly River: Cholera and Cover-Up in Post-Earthquake Haiti.* London: ILR Press, 2016.

Lutgens, Frederick K. and Edward J. Tarbuck. *Foundations of Earth Science*, 7th ed. New York: Pearson, 2014.

Markley, Robert. "Monsoon Cultures: Climate and Acculturation in Alexander Hamilton's *A New Account of the East Indies.*" *New Literary History*, vol. 38 (2007), pp. 527–550.

Oishi, Shigehiro. "Socioecological Psychology." *Annual Review of Psychology*, vol. 65 (2014), pp. 581–609.

Prager, Ellen. *Furious Earth: The Science and Nature of Earthquakes, Volcanoes, and Tsunamis.* New York: McGraw-Hill, 2000.

Reilly, Benjamin. *Disaster and Human History: Case Studies in Nature, Society, and Catastrophe*, 2nd ed. Boone, NC: McFarland, 2022.

Ritchie, David. *The Encyclopedia of Earthquakes and Volcanoes.* New York: Facts on File, 1994.

Vandello, Joseph A. and Dov Cohen. "Patterns of Individualism and Collectivism Across the United States." *Journal of Personality and Social Psychology*, vol. 77, no. 2 (1999), pp. 279–292.

Walder, Deborah J. *et al.* "Prenatal Maternal Stress Predicts Autism Traits in 6/12 Year-Old Children: Project Ice Storm." *Psychiatry Research*, vol. 219 (2014), pp. 353–360.

Works Cited

Bankoff, Greg. *Cultures of Disaster: Society and Natural Hazard in the Philippines.* New York: Routledge Curzon, 2003.

Cohn, Samuel K. *Epidemics: Hate and Compassion from the Plague of Athens to AIDS.* New York: Oxford University Press, 2018.

Courtwright, David T. *Violent Land: Single Men and Social Disorder from the Frontier to the Inner City.* Cambridge, MA: Harvard University Press, 1996.

Curtis, Valerie, Mícheál de Barra, and Robert Augner. "Disgust as an Adaptive System for Disease Avoidance Behavior." *Philosophical Transactions of the Royal Society of London, B*, vol. 366 (2011), pp. 389–401.

Del Guidice, Marco, Steven W. Gangestad, and Hillard S. Kaplan. "Life History Theory and Evolutionary Psychology." In David M. Buss (Ed.), *The Handbook of Evolutionary Psychology*, vol. 1. Hoboken, NJ: Wiley, 2016, pp. 88–114.

Fincher, Corey L., Randy Thornhill, Damien R. Murray, and Mark Schaller. "Pathogen Prevalence Predicts Human Cross-Cultural Variability in Individualism/Collectivism." *Proceedings: Biological Sciences*, vol. 275 (2008), pp. 1279–1285.

Gelfand, Michele *et al.* "Differences Between Tight and Loose Cultures: A 33-Nation Study." *Science*, vol. 332 (2011), pp. 1100–1104.

Harrington, Jesse R. and Michele Gelfand. "Tightness-Looseness Across the 50 United States." *PNAS*, vol. 111, no. 22 (2014), pp. 7990–7995.

Kruger, Daniel J., Heitor B. F. Fernandez, Suzanne Cupal, and Gregory G. Homish. "Life History Variation and the Preparedness Paradox." *Evolutionary Behavioral Sciences*, vol. 13, no. 3 (2019), pp. 242–253.

Lee, Ju-Yeon, Sung-Wan Kim, and Jae-Min Kim. "The Impact of Community Disaster Trauma: A Focus on Emerging Research of PTSD and Other Mental Health Outcomes." *Chonnam Medical Journal*, vol. 52, no. 2 (2020), pp. 99–107.

McNeill, William H. *Plagues and Peoples.* New York: Anchor Books, 1977.

Nesse, Randolph M. "Evolutionary Psychology and Mental Health." In David M. Buss (Ed.), *The Handbook of Evolutionary Psychology*, vol. 2. Hoboken, NJ: Wiley, 2016, pp. 1007–1026.

Oishi, Shigehiro and Asuka Komiya. "Natural Disaster Risk and Collectivism." *Journal of Cross-Cultural Psychology*, vol. 48, no. 8 (2017), pp. 1263–1270.

Schaller, Mark. "The Behavioral Immune System and the Psychology of Human Sociality." *Philosophical Transactions of the Royal Society of London, B*, vol. 366 (2011), pp. 3418–3426.

Schaller, Mark. "The Behavioral Immune System." In David M. Buss (Ed.), *The Handbook of Evolutionary Psychology*, vol. 1. Hoboken, NJ: Wiley, 2016, pp. 206–224.

World Health Organization. "COVID-19 Pandemic Triggers 25% Increase in Prevalence of Anxiety and Depression Worldwide." March 2, 2022. Available at: https://www.who.int/news/item/02-03-2022-covid-19-pandemic-triggers-25-increase-in-prevalence-of-anxiety-and-depression-worldwide

2 Earthquakes

On September 1, 1923, journalist W. Reer Harris stepped out of the Tokyo office of the Kokusai News Service just before noon and set off for Tokyo's main street, the Ginza, where he hoped to get a bite of lunch. The air was warm and muggy, as was typical of a Tokyo September, but Harris might have noted that a breeze was beginning to pick up, bringing some relief from the heat. As he walked, Harris probably kept an ear open for the cracking call of the don, an antique cannon fired from the palace plaza every day to mark high noon.

On that day, however, the don was never fired. Less than two minutes before 12 noon, Tokyo was struck by the first of a series of earthquake waves so severe they disabled all the seismographs at Tokyo's Central Weather Bureau. As Harris passed through Tokyo's Hibiya Park near the Imperial Palace, the earth suddenly "seemed to rock," nearly throwing Harris to the ground. Clutching a lamppost for support, Harris watched in terror as the trees around him swayed and gyrated. Fearing that he might be struck by a falling limb, Harris "ran for [his] life" for the open center of park as soon as the shaking seemed to subside, only to be knocked off his feet by a second and perhaps even more powerful shock. "It was the most peculiar sensation," Harris remembered; "the earth was heavily pulsing internally, as if possessed by some demon." Harris watched as the buildings on the park's edge swayed and shuddered, including one large three-story restaurant that "collapsed and tumbled down like a pack of cards." At this point Harris was joined in the Hibiya Park by hundreds of Tokyo residents, who threw themselves into the park's open space to escape the falling timbers and masonry.

Once the "earth became quiescent" once again, Harris hurried back to his office to check on his colleagues, only to find that the office had been damaged along with the rest of the neighborhood. "Pillars and walls [were] bulging out," Harris recalled, "and the pavement and road [were] littered with fallen roof-tiles and other debris." But the news agency itself was lucky – none of his colleagues, thank God, had been injured during the quake. Relieved, Harris headed back over to the Ginza – not as a

DOI: 10.4324/9781003436805-3

hungry man looking for lunch, but as a reporter looking for news. There, he found to his dismay that "some of the buildings had collapsed, others partially fallen, and the rest [were] badly shattered, with piles of debris in the streets – a scene of devastation hard to describe."

At least the disaster was over. Or was it? A pall of smoke and the smell of burning timbers began to fill the air, as fires sprang up in several parts of the city with "alarming rapidity" – Harris would later claim that "in places the flames leaped from house to house quicker than one could walk." Harris may or may not have realized it at the time, but the real horror of the Great Kantō Earthquake and firestorm had only just begun.

The Science of Earthquakes

Earthquake Origins

In technical terms, anything that creates seismic waves is an earthquake, including man-made events like mine collapses and underground nuclear detonations. Artificial lakes created by dams can also create quakes because the weight of the water puts new pressures on existing fault lines, occasionally leading to mild tremors. Earthquakes can also be spawned by the deliberate pumping of fluids into the ground, as epitomized by a bizarre 1962 incident near Denver, Colorado, where earthquakes detectible by nearby seismographs were triggered when the U.S. military disposed of thousands of tons of liquid nerve gas by injecting it deep into the earth's crust. More recently, earthquakes have been linked to "fracking" or hydraulic fracturing in the United States, which involves pumping fluids deep into shale rock layers underground to release natural gas. In Oklahoma, where fracking is common, the number of detectable tremors increased 900-fold between 2008 and 2015. One such quake, striking on November 6, 2011, was powerful enough to buckle U.S. Route 62 in three different locations and to cause serious damage to homes in the Oklahoma town of Prague. This quake helped to convince Oklahoma to pass new legislation limiting underground wastewater pumping, but in the short run this will do little to stop the problem, since water already pumped underground is expected to seep downwards and trigger new quakes for years into the future.

Despite these outlying examples of man-made tremors, earthquakes are generally the result of natural processes. Plate tectonics is the most usual culprit. Convection in the mantle ensures that both oceanic and continental plates shift their positions over time, colliding, separating, and rubbing alongside each other in the process (Figure 2.1). Mind you, the movement of these plates is extremely slow. The slowest plates poke along at little more than a centimeter annually, while a "fast" plate like the Juan de Fuca of the Pacific Northwest hurries along at the comparatively rapid rate of 16 centimeters a year, about the same speed as a growing strand of

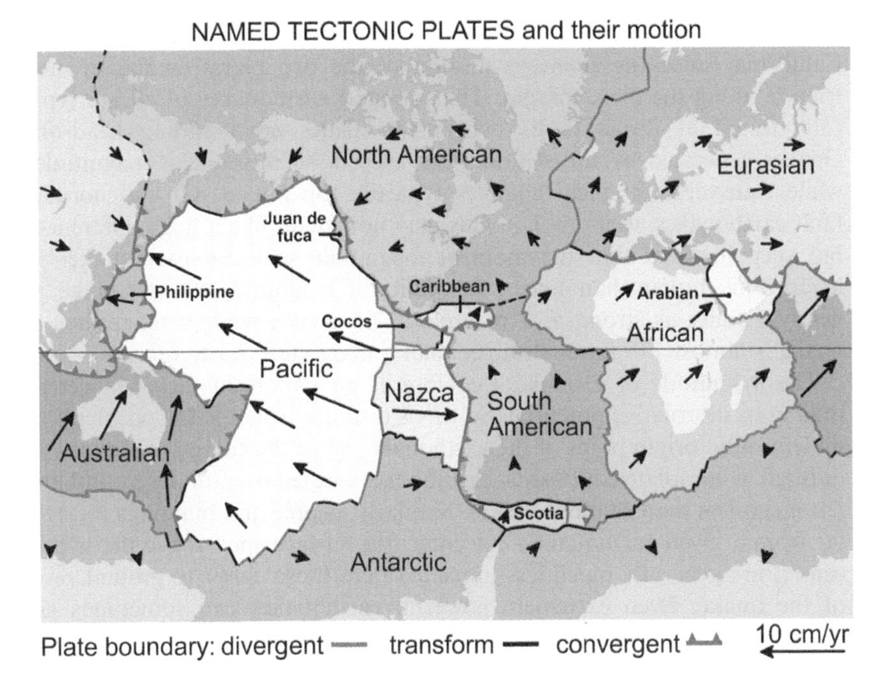

Figure 2.1 Relative movement of the world's tectonic plates
Source: Wikimedia Commons, created by Hughrance.

human hair. Nonetheless, given the staggering weight of the plates themselves, which can be up to 100 kilometers thick, the cumulative kinetic energy involved in plate movement is enormous.

Unfortunately for human beings living near plate margins, this kinetic energy is not released gradually, but in sudden spurts. Most of the time, tectonic plates are prevented from movement by friction with neighboring plates. Unable to shift location, the plates deform instead, much like a steel spring changes shape when compressed. Plate movement only occurs when the potential energy stored within the plates exceeds the frictional energy holding them in place, at which point the spring is sprung and the plate moves abruptly. If two or more plates have been locked together for a long time, the resulting movement can be dramatic. During the Sumatra-Andaman earthquake of 2004, for example, the sudden release of 700 years of stored kinetic energy caused the Indian Ocean and Burman plates to move up to 50 feet relative to each other in a matter of minutes along a front of over 1200 kilometers, releasing the energy equivalent of 1,500 Hiroshima-type nuclear bombs in the process.

Not all plate collisions are able to generate such colossal forces, of course. In normal faults, where the plates are moving away from each other, the friction is minimal and pressure is released constantly, leading to frequent but low-intensity earthquakes. More energy can build up in transverse or slip-strike faults, such as the famous San Andreas Fault in California, since the sideways motion of the two plates creates greater friction along the plate margin. The strongest earthquakes of all are typically found at thrust faults, where two plates are colliding head-on. Quakes generated by thrust faults can exceed 9.0 in moment magnitude, while transverse fault earthquakes typically top out at 8.0 and normal fault earthquakes at about 7.5. This may not sound like a huge difference, but keep in mind that the moment magnitude scale for earthquakes is logarithmic rather than linear. In reality, a magnitude 8 earthquake is nearly 6 times as strong as a magnitude 7.5 quake, while a magnitude 9 quake is almost 178 times stronger than a magnitude 7.5 tremor.

The magnitude of a quake, therefore, is an important factor in determining its destructive potential. So too is distance from the hypocenter, the underground origin point of the earthquake. Since the energy of a quake is diffused as it radiates outwards, earthquakes located deep underground are felt much less than shallow quakes. Similarly, people and buildings located far from a given earthquake's epicenter (the surface spot above the hypocenter) are generally much less impacted than those closer to ground zero of the quake. Even extremely powerful earthquakes can sometimes go almost entirely unnoticed if they strike a depopulated area – like the proverbial tree in the forest, they can fall without making a sound. Conversely, a moderate quake can wreak horrific death and destruction if the hypocenter is shallow and close to a vulnerable population. Case in point is the Haiti Earthquake of 2010, which killed at least 85,000 people despite having a moment magnitude of only 7.0, since the hypocenter was just 15 kilometers below the extremely vulnerable population of Haiti's overcrowded and impoverished capital, Port-au-Prince.

While most earthquakes, including the Sumatran-Andaman and Haitian quakes, occur along the boundaries of tectonic plates, the occasional earthquake does strike outside of these zones. Volcanoes, for example, produce low intensity earthquakes during the process of eruption, and, as we will see in Chapter 3, some "hot spot" volcanoes are located nowhere near a plate edge. In addition, the occasional earthquake strikes the interior of a tectonic plate, usually due crustal movement along ancient and long-inactive fault lines. One such fault line is the New Madrid Seismic Zone, an ancient and mostly inactive rift system that helps determine the course of the Mississippi River and which erupted into a series of powerful magnitude 7+ earthquakes from December 1811 to February 1812. These quakes were strong enough to rearrange furniture, crack plaster, and ring church bells in East Coast American cities; closer to the epicenter, they flattened houses and appeared to observers to temporarily reverse the flow of the Mississippi River. Luckily, death and property damage were minor in this underpopulated and underdeveloped region of the newborn United States. Should a similar set of quakes strike the same region today, which is now much more densely inhabited, the results would likely be dire. It should be noted that, in terms of property damage at least, intraplate earthquakes such as the New Madrid quakes tend to punch well above their normal weight class. Precisely because they are so rare, these quakes shake cities and towns where earthquake-resistant structures are uncommon, meaning they inflict damage out of proportion to their actual strength.

Earthquake Hazards

Whether they strike on the edge of a tectonic plate or its center, earthquakes do their damage through various types of seismic waves, including primary waves, secondary waves, and surface waves. Quick-moving and generally non-destructive P-waves (also known as pressure waves) are the first seismic wave to arrive, and engineers have capitalized on this fact by designing devices like electrical grids and elevator systems to shut down automatically when P-waves are detected. Next on the scene are the S-waves, oscillating waves radiating out from the quake center. Since these waves cause structures to sway from side to side, they often account for the lion's share of the damage done to normal buildings, which are typically designed to withstand the vertical force of gravity but not the horizontal stresses that these earthquake waves induce. The last to arrive are the surface waves, low-frequency vibrations that are particularly damaging to taller buildings like skyscrapers. These latter waves also conduct energy more efficiently over long distances, explaining the odd phenomenon sometimes seen in major earthquakes when multi-story apartment buildings far from the epicenter collapse in a quake while smaller surrounding structures in the same city seem to be entirely unaffected.

The impact of these waves at any given location is also influenced by the underlying geographical and geological characteristics. Buildings constructed atop solid rock tend to be fairly resistant to earthquakes. Conversely, neighborhoods constructed on softer soil, such as river floodplains and former agricultural land, often suffer disproportionately. During the major tremor that shook Mexico City in 1985, for example, the worst of the damage occurred in neighborhoods built atop the bed of old lake Texcoco, where (according to Wisner et al., 2004) the soft underlying alluvial deposits were up to 60 meters deep. Such soils shake like gelatine when subjected to seismic shocks or even experience liquefaction, temporarily taking on the consistency of quicksand.

Soil liquefaction is a particular problem in reclaimed or fill land. In some coastal cities, geographical constraints or human greed has driven the construction of artificial land atop marshes, flood plains, or coastal margins. However, adding unconsolidated fill material to unstable soils is a literal recipe for disaster. The city of Messina in Sicily, for example, suffered a catastrophic quake in 1908 in part because much of the downtown area was built upon earlier layers of fill – some of which was the rubble of buildings that had been demolished by previous earthquakes! As we shall see below, San Francisco committed this same sin after its own 1906 quake when it used debris from the quake to expand its own bayfront territory, likely laying down the foundations of future earthquake disasters in the process. These catastrophic quakes, when they occur, will underscore the claim by Oliver-Smith (2012) and others that disasters are not one-off events but rather processes with deep historical roots.

So how do earthquakes kill? Seismic waves are actually more disagreeable than dangerous to people out in the open, as we saw in the tale of W. Reer Harris that began this chapter. Rather, deaths during earthquakes are almost always caused by the collapse or combustion of man-made structures. Quakes kill by pinning us under falling beams, bludgeoning us with bricks from toppled walls, or braining us with plummeting roof tiles. Those who survive the initial collapse of their homes often fall victim to urban fires, which thrive after quakes due to upended candles, collapsed fireplaces, gas line leaks, and electrical shorts. Earthquakes therefore kill mainly by turning our own living environment against us. As seismologist W. H. Hobbes wisely observed after the Messina quake, "if an army in tents had been camped on the site of Messina on the morning of the 28th of December last, the loss of life and property would have been insignificant." But tents were out of fashion in Messina, where most people inhabited multi-story buildings with thick but fragile masonry walls and unsupported floors and ceilings. These "man trap" buildings pancaked during the quake, adding considerably to Messina's huge death toll, which may have reached 100,000. As we will see in the case studies later in this chapter, this is the rule rather than the exception: the death and destruction "caused" by earthquakes are almost always of our own making.

Earthquakes hazards, therefore, are not necessarily disasters but become so when they interact with specific human vulnerabilities. This chapter has already discussed several ways in which people increase the likelihood of disaster, such as living on fill land or in unsupported masonry buildings that are susceptible to seismic shocks. Houses constructed precariously on steep hillsides are especially vulnerable to earthquakes because of the landslide risk, as are homes and buildings that are not being properly maintained. In addition, houses not built to minimum construction standards and/or constructed with inferior materials are much less likely to withstand earthquake tremors.

The common link between these sources of earthquake vulnerability, of course, is poverty. Economic disadvantage means the poor are far more likely to occupy substandard housing on marginal lands. Impoverished neighborhoods also tend to have less access to health services and often lack the transportation infrastructure needed to provide quick relief to a stricken area. Since poor populations also lack political clout, they generally receive far less than a fair share of the post-disaster aid. Worst of all, the destruction wreaked by an earthquake serves only to deepen the poverty of this worst-affected population, and thus (in the words of Wisner et al., 2011) "further marginaliz[es] those who were already living at the margin before the event." Indeed, As Wisner et al. pointed out in 2004, some earthquakes are so selective in who they harm they might better be called "class-quakes." Case in point was the 1976 Guatemala Earthquake, which claimed most of its 22,000 victims in destitute mountainous districts and in urban squatter settlements but left the "upper and middle classes virtually unscathed."

That being said, supposedly class-free communist states have suffered their own share of horrific earthquake disasters, since such states suffer from many of the same vulnerabilities as capitalist societies plus others of their own making. In 1976, for example, a shallow 7.6 magnitude quake triggered the worst earthquake disaster of the twentieth century when it struck the Chinese coal-mining city of Tangshan, killing as many as 300,000 people, fully one-quarter to one-fifth of the pre-quake population. Much of the fault for the Tangshan quake's high death toll lies with China's communist government, which wanted to close the industrialization gap between China and the West as quickly as possible and thus pushed for the rapid development of Tangshan and its coal deposits while underestimating the region's seismic risks. As a result, Tangshan at the time of the quake was a city of unreinforced brick structures and quickly constructed, jerry-built rural homes, both of which were highly susceptible to lateral shaking during an earthquake. Worse yet, since Tangshan lies in northern China, heavy roofs capable of withstanding winter snows were the norm. During the earthquake these roofs collapsed down upon Tangshan's inhabitants like "a great hand slapping the earth," in the words of Tangshan quake historian James Palmer (2012). Others died when heaps of slag toppled over during the quake, burying entire neighborhoods beneath tons of industrial debris.

Unfortunately for Tangshan, the communist state's recovery efforts prioritized national strategic and political goals rather than the well-being of the survivors. Offers of assistance from the outside world were rebuffed; as Palmer points out, given the virulently anti-Western political atmosphere of Mao's ongoing Cultural Revolution in China, agreeing to accept aid from "Foreign Devils" would have been sacrilege. China's People's Liberation Army did descend on Tangshan shortly after the quake to pull the living and dead from the building ruins, and large amounts of food were shipped to the disaster zone. However, very little assistance reached the badly stricken rural communities outside of Tangshan, as the state's priority was to get the city's coal mines back on-line again as soon as possible. For the same reason, few resources were invested in rebuilding housing or restoring the living standards of Tangshan's quake survivors. As a result, the city resembled a gigantic refugee camp for years after the Tangshan quake, and the work of rebuilding the city would not begin until 1979.

Of course, human beings are by no means powerless in the face of earthquake hazards. Tall buildings can be braced for lateral support, helping them to withstand the side-to-side shaking of S-waves. Large structures can also be engineered to rest on a floating base, so that when the quake strikes, the foundation moves but the building itself does not. Some buildings in Japan are not attached to the ground at all, but instead hover atop a column of air maintained by a compressor system. Buildings can also be constructed of lighter and more flexible materials, such as wood. As a precaution against inevitable post-quake fires, cities can be zoned so that neighborhoods of wooden houses are separated from each other by firebreaks, such as rows of concrete buildings or urban parks. What is more, adaptations to the threat of earthquakes are by no means limited to the modern world. As the next section will illustrate, earthquake awareness and preparedness were often built into the "cultures of disaster" of seismically-prone world regions.

Earthquake hazards may also help to shape cultures of disasters through evolved human mental mechanisms, including but not limited to the behavioral immune system discussed in Chapter 1. In one fascinating recent study of the link between human religiosity and the prevalence of natural disasters, for example, Jeanet Sinding Bentzen (2019) found a strong link between human religiosity and natural disaster frequency, particularly earthquakes and tsunamis. What these two disasters have in common, Bentzen contends, is their unpredictability, making them difficult to prepare against. In contrast, volcanic eruptions have a more modest impact on religiosity, and tropical storms have almost no impact, presumably because such storms are expected almost annually and arrive in predictable seasons of the year.

Bentzen suspects that the close link between quakes and religiosity arises from a psychological mechanism known as the religious coping hypothesis. In the face of adversity, this theory holds, people will "draw on religious practices to understand and deal with unbearable and unpredictable situations" – in these circumstances, religion offers both a means of understanding adversity, and possibly a mechanism to prevent its recurrence. Nonetheless, while religious culture is important, Bentzen suggests that it plays a subordinate role to universal human psychology. People of all world regions and all world religions (except Buddhism) respond with the same spike in religiosity after a quake disaster, and this effect was robust even when cross-referenced with education and socioeconomic factors. What is more, while much of impact of quakes on the religiosity of individual humans fades after 6–12 years, she found that a "residual impact" remains at the community level even after that period that can be "transmitted across generations." If Bentzen is correct, a quake-prone society's "culture of disaster" might derive in part from specific human psychological mechanisms.

On a shorter-term scale, the disease cues that accompany major earthquakes can trigger the behavioral immune system, often leading to exclusionary behavior or even xenophobic violence against perceived social outsiders in the aftermath of the quake. The accusations that the Jews of Europe triggered the Black Death in Europe are well known. Less well known is that Jews were also blamed for earthquakes, in particular the devastating 1348 Friuli quake in Northern Italy, which led to a large number of the region's Jews being burned alive. Nor did this link between earthquakes and xenophobia end with the Middle Ages. In all four of the case studies below, outsiders were singled out for persecution in the aftermath of a quake, and in two (arguably three) of these examples, this xenophobic impulse led to bloodshed – or even massacre, in the case of the Great Kantō Earthquake. In suggesting that these xenophobic outbreaks had a common psychological origin, I am not claiming that other factors, such as economic competition, political rivalry, and cultural identity, played no role in the outcome of these events. Culture in fact played a crucial role, since what it means to be an "outsider" in any given society is as much a cultural construction as a self-evident fact. Still, the occurrence of similar episodes of exclusionary violence in quite different societies demands some sort of explanation, and the behavioral immune system theory provides the best one currently available.

Earthquake Cultures

In many traditional cultures, the spasmodic twitching of the earth during an earthquake seemed to demand an animalistic explanation. Hindus imagined that the god Vishnu, in his role as world protector, rested upon a world snake, which was coiled upon the back of a turtle, which in turn was balanced upon four enormous elephants whose movements shook the earth. By the eighteenth century some Japanese ascribed earthquakes to the floundering of *Namazu*, an enormous catfish beneath the earth that was normally held in place by a "cap stone" but who would flop and thrash about if the stone was dislodged (Figure 2.2). Interestingly enough, this linkage between catfish and earthquakes seemed to be confirmed in a series of experiments in 1930s Japan, which established that the behavior of aquarium catfish could be used to predict imminent earthquakes with 80% accuracy.

In the West, there has long been a sharp divide over how to interpret earthquakes, with erudite and philosophical perspectives on the natural origins of earthquakes co-existing with popular perspectives that focused on divine causes. From the sixth century BCE until the late Middle Ages, most educated men ascribed to Aristotle's opinion that earthquakes were caused by the release of compressed air from underground caves. Earthquakes obeyed the laws of nature, Roman scholar Lucretius believed, and he rejected the notion that quakes were divine punishments for our earthly sins. In the popular mind, however, the link between earthquakes and supernatural punishment was self-evident. Greeks and Romans ascribed earthquakes to the anger of Poseidon (Neptune to the Romans), perhaps

Figure 2.2 Namazu, the giant catfish blamed for Japanese earthquakes
Source: University of British Columbia Library, Rare Books and Special Collections.

because Poseidon was the god of the sea and earthquakes in the Mediterranean were frequently accompanied by tsunamis. This connection between quakes and divine anger became even stronger in the Christian era. The Jewish God of the Old Testament, like Poseidon before him, made his anger against mankind manifest through earthquakes. In the New Testament, Jesus's death unleashed an earthquake that shook the temple in Jerusalem. What is more, both Luke and Mark predicted that quakes would herald Jesus's return, and the Book of Revelations promised that "a great earthquake such as there has never been since man was on the earth" would mark the end of days.

According to Rienk Vermij (1998; 2020), these various early views about the causes of earthquakes changed over time to fit with prevailing European cultural beliefs. Pre-modern European scholars placed earthquakes within a general category of "meteors," a grab-bag of unusual natural events that also included freak storms, comets, and volcanic eruptions. Renaissance scholars studying such meteors generally adopted a "philosophical, largely secular understanding of nature" they inherited from classical philosophy. However, this attitude changed markedly during the religiously charged sixteenth century, the era of the Reformation and Counter-Reformation. During this period, elite and popular perspectives on earthquakes moved closer together, with Protestants and Catholics alike shifting toward interpreting natural disasters, including earthquakes, as scourges of God and signs of the wrath of the Almighty. By the era of the Scientific Revolution and the Enlightenment, however, this fire-and-brimstone interpretation of earthquakes had begun to fade, at least in educated circles, replaced by renewed attempts to find rational explanations for these events. Earthquakes were now explained variously as underground movements of gases, chemical reactions between iron and sulfur, or even by "subterranean thunderstorms" arising from "electrical fluid" within the earth. We will return to this tension between religious and more secular interpretations for earthquakes later in this chapter when discussing the Lisbon Earthquake of 1755, which struck during a transitional period between religious and natural science understandings of earthquake disasters.

While the philosophers and theologians debated the meaning of earthquakes, European populations in seismically-active areas took practical steps to mitigate earthquake hazards as part of their day-to-day culture of disaster. According to Greg Bankoff (2015), Byzantine buildings from the third century CE onwards incorporated *hatillar* reinforcement beams as earthquake shock absorbers, and this practice was incorporated into Ottoman architecture after the conquest of Constantinople. In earthquake-prone Italy, urban planners tried to incorporate wide streets and urban plazas into their designs to provide safe places away from falling masonry. Southern European builders from the fourteenth century onwards also experimented with incorporating a cage-like timber frame called the *casa baraccata* (Italian for "wooden barracks") into the interior of masonry buildings to support the walls and ceilings in case of a major quake.

European populations also sought religious protection against earthquakes, including appeals to Emygdius, the patron saint of earthquakes, who is often depicted in religious art holding a bracing hand against a crumbling wall. In the aftermath of quakes, stricken communities might organize penitential processions; Jussi Hanska (2002), for example, reports that the Christian community of Tours was so terrified by a 1579 earthquake that three hundred of them marched barefoot in the wintertime to make atonement for their sins. Good Christians also fled to churches during earthquakes, hoping for protection from calamity within this sacred space. Unfortunately, seeking shelter within heavy masonry structures during an earthquake is not a great idea – during the 1348 Friuli earthquakes all five hundred people who had sought protection within the parish church in the Austrian town of Villach were killed when that church collapsed.

In the meantime, a quite different set of belief about earthquakes had developed halfway around the world in early modern Japan. As with Europe, there was a distinction between popular and erudite beliefs about quakes. In art and popular storytelling, earthquakes were associated with the movement of a giant catfish beneath the earth, but Gregory Smits (2012) argues that learned Japanese considered the catfish more as a symbol of earthquakes than a direct cause. Rather, for elite Japanese whose worldview was informed in large part by yin-yang theory, earthquakes were explicable as the result of pent-up yang energy that had been restrained too long by yin. The Japanese also understood (perhaps better than Europeans) that earthquakes had a focus and that energy radiated outwards from this center point. Another interesting difference in European and Japanese understandings of earthquakes is that earthquakes were not regarded universally as a calamity in Japan. Past earthquakes had brought benefits to some Japanese, for example, by uplifting coastal land and turning former tidal flats into exploitable real estate. Earthquakes also served to redistribute wealth from the rich to the poor, since the rich were more likely to own the property destroyed by earthquakes, while poor craftsmen and manual laborers benefited from post-quake demand for their services and the consequent rise in wages. For this reason, one popular print depicting the Ansei Edo Earthquake of 1855 re-imagines the catfish *Namazu* as a whale with a stream of gold coins spouting from his blowhole.

Despite these differences, Japan's culture of earthquakes was similar to that of Europe in one respect: there was a tendency in both Europe and Japan to interpret quakes as signs of divine displeasure. By the nineteenth century, at least, many Japanese saw earthquakes not as random occurrences but as *yanaoshi*, or "world rectification" events, which (in the words of Smits, 2006) "literally shook up a society that had grown complacent, imbalanced, and sick." Smits argues in fact that the Ansei Edo quake was interpreted by many Japanese, who were already upset about Commodore Matthew Perry's 1853 expedition to Japan, as a further sign that the ruling *bukufu* military government had lost its legitimacy. This impression was reinforced by the liquification-induced destruction of the swampy islands in Tokyo's harbor upon which the *bukufu* had positioned coastal batteries, guns that were visual symbols of the shogun's power. Smits therefore links the quake to the restoration of the Japanese imperial government, arguing that, while the quake itself was not a revolutionary event, it nonetheless helped to prepare Japan for profound future changes.

Lisbon, 1755

Almost exactly one hundred years before the Ansei Edo quake, the Portuguese city of Lisbon was shaken by a "world rectification" event of its own. According to Ryan Hagan (2021), the Lisbon quake "shocked Western Civilization more than any other event since the fall of Rome," reignited Europe's interest in tectonic processes, and altered the trajectory of the ongoing Enlightenment intellectual movement. The sheer scale of the disaster also horrified contemporary Europeans, to the point that the name of Lisbon acquired "connotations of evil similar to the word Auschwitz today" (Paice, 2008).

Lisbon Before the Quake

At the time of the quake, Lisbon was a city of up to 200,000 souls and was one of Europe's wealthiest capitals, largely due to the vast revenues generated by Portugal's sprawling overseas empire. Imports of sugar, tobacco, and gold from Brazil helped make Lisbon the third biggest port in the world, and the revenues from this trade funded lavish lifestyles in the Portuguese capital. Pre-quake Lisbon, T. D. Kendrick (1955) claimed, was "staggeringly rich ... aloof, proud, happy, [and] spendthrift." This wealth was on display in what Nicholas Shrady (2008) called Lisbon's "landscape of devotion," – the city's superabundance of churches and abbeys, all built using architectural styles copied from Rome. At the time of the quake, in fact, as many as 10% of Lisbon's citizens belonged to a holy order, one of the highest rates in Europe.

Despite this building spree, much of Lisbon still retained a "warren-like topography" inherited from the medieval era. While the rich lived at higher elevations on Lisbon's margins, most of the poor lived in the downtown Baixa region, where the streets were extremely narrow and seemed to run in random directions. Many houses in the Baixa were flimsy wooden constructions built far closer together than was wise in a seismically-active region. And the Portuguese should have known better; Lisbon had been shaken by major earthquakes before, in 1321 and then again in 1531. By 1755, unfortunately these previous quakes had been all but forgotten.

Underlying geology compounded the earthquake vulnerabilities inherent in Lisbon's medieval building plan. Lisbon was founded where the Tagus River estuary meets the Atlantic Ocean, and, as a result, much of the city lies atop coastal dunes and silt or else a weak mixture of sand and sandstone. Making matters worse, Lisbon's Baixa neighborhood was built atop alluvial deposits laid down by two creeks that originally emptied into the Tagus, and this already-weak core zone was made still weaker by new boom-time construction along the coast on "reclaimed" fill land. Soils of this sort are particularly vulnerable to deformation and liquefaction when impacted by seismic waves. Worst of all, a tectonic fault line passes directly under Lisbon itself, though in the case of the 1755 earthquake, this particular fault was not responsible for the destruction.

Lisbon's weak soils hosted one of Europe's most diverse urban populations, both in terms of ethnicity and religion. Many Spaniards lived in Lisbon, including a troublesome contingent of deserters from the Spanish

Army. Lisbon's mercantile opportunities attracted large numbers of expatriates elsewhere in Catholic Europe as well, including Irishmen, Frenchmen, and Poles. Despite Lisbon's reputation as a Catholic bastion, Protestant merchants were a common sight on Lisbon's streets, especially the British, who dominated Lisbon's foreign trade. Lisbon also reflected Portugal's own internal diversity, much of which it inherited from the period of Islamic hegemony in the Iberian Peninsula. Many of Lisbon's citizens were so-called "New Christians," the descendants of Jews and Muslims who had been forced to convert after the conquest of Granada in 1492. These *Conversos* (former Jews) and *Moriscos* (former Muslims) were widely distrusted in Portugal and were treated as second-class citizens, excluded from power and high social circles since they lacked *limpieza de sangre* (cleanliness or purity of blood). Distrust of the new Christians was exacerbated by Lisbon's wide-ranging economic ties; the fact that many Portuguese "New Christians" had historically reverted to their old religious identity once they left Portugal contributed to distrust of the New Christians who remained.

The Lisbon Quake: Fire and Water

On November 1, 1755, these underlying social tensions were shaken by the sudden release of seismic tension along the boundary between the African and Eurasian tectonic plates. Although the exact epicenter location is unknown and subject to some debate, it was probably along a thrust fault about 300 kilometers to the southwest of Lisbon beneath the Atlantic Ocean. Estimates of the moment magnitude range widely, from 7.7 to as high as 9.1; if the latter is correct, it was probably one of the strongest earthquakes to strike Europe in recorded history. The severity of this quake is reflected in the size and range of the tsunami wave that it generated, which would reach 20 meters in height in some parts of North Africa, 3.5 meters across the Atlantic in distant Barbados, and may even have caused some damage as far away as Ireland and Brazil. The earthquake itself was felt appreciably as far away as Greenland, and in northern Germany it set some chandeliers swinging like pendulums, much to the puzzlement of church-goers.

Closer to Lisbon, the oncoming seismic waves were first noted by ships in the harbor; one English sailor reported feeling an "uncommon motion" and thought that his ship, despite being in the open water, had somehow run aground. On land, the earthquake began with a "horrible subterranean noise," variously described like the clatter of carriages in a busy street, distant drums, or even a "hollow, distant rumbling of thunder." Soon after, the city itself began to sway and buckle with "a tumbling sort of motion … like waves in the sea." Buildings began to grind against each other with "the most dreadful jumbling noise" and many collapsed, crushing those within and threatening those nearby with falling rubble.

The main quake consisted of three distinct pulses that lasted in total about 10 minutes, reducing the city's structures to "one confus'd heap of rubbish," in the words of English quake survivor Thomas Chase. As is typical of earthquakes, larger masonry structures proved to be particularly vulnerable to the lateral shaking of the quake, though liquefaction of the soft soils of the Baixa contributed to widespread damage even among more resilient timber-built homes.

Desperate to avoid falling masonry from quake-shattered buildings, many of Lisbon's inhabitants fled to the relatively open space of the waterfront for safety. This proved to be a fatal mistake. About 30 minutes after the main quake, a 6-meter tsunami surged up the Tagus River and into Lisbon's harbor, where it "smashed anchors, broke cables, and then entered the streets and squares near the riverbank." After penetrating several blocks inland, the first tsunami sloshed back out to sea, carrying hundreds of Lisbon's residents with as it withdrew, only to surge back again twice more with the same foaming fury. Together, these waves transformed the normally safe waters of Lisbon's harbor into "a confusing mass of entangled ships' masts … a horrible cemetery of cadavers." Ultimately, the tsunamis triggered by the Lisbon quake may have claimed 3,000 lives, and this death toll was likely exacerbated by a cruel coincidence: the tsunami arrived in the Tagus at close to high tide.

Lisbon's destruction was made complete by another unfortunate coincidence, the fact the quake struck on November 1, All Saints Day, an important holiday in the Catholic liturgical calendar. As a result, when the earthquake waves arrived, "every altar and every chapel … was illuminated by a number of wax tapers and lamps." Toppled by the earthquake, these lamps and candles kindled fires that leapt quickly through the shattered city and did not begin to subside until fully five days after the quake. At their height, these fires generated a "firestorm" – a small-scale convection cell, in which superheated air above an inferno rises into the atmosphere and is replaced by surface-level air from the surrounding area, feeding more oxygen to the flames in the process. Once started, this self-strengthening process only ends when it runs out of fuel, which in this case was the fallen beams, door and window frames, and other timbers of quake-shattered Lisbon. The firestorm proved to be the killing blow for many of Lisbon's residents, especially old or crippled Lisboners and those who had survived the initial quake but were now pinned under rubble or trapped beneath collapsed buildings.

Taken together, the combined forces of quake, tsunami, and firestorm wrecked almost unimaginable havoc on the city of Lisbon. All of Lisbon's 40 churches were damaged and 35 were outright destroyed, as were 65 of Lisbon's 75 convents and most of Lisbon's palaces. Damage was almost total in the low-lying Baixa district. Contemporary observers described the former site of Lisbon as little more than a "chaos of stones," and the combined losses in real estate, warehoused goods, and personal property were probably equal to over half of Portugal's annual GDP. In terms of human casualties, Mark Molesky (2015) estimated that up to 40,000 may have died in Lisbon, including up to 25,000 killed by the quake itself, another 7,000 lost to the firestorm, 3,000 drowned or swept away by the tsunamis, and 3,000 or more who died later of injuries sustained during the disaster. Molesky claims the death toll among women was disproportionately higher than among men, largely as men were better able to extricate themselves from the post-quake wreckage and to resist being pulled to sea by the outflowing tsunami. What is more, as is typical of such disasters, commoners suffered worse than aristocrats, in part because the rich tended to live on the peripheral heights of the city rather than in the inner city, and in part because many aristocrats were not even in town when the quake struck, preferring to celebrate the holiday at private chapels on their own rural estates.

Aftermath and Legacy

In the days after the earthquake, Lisbon's survivors exhibited a set of behaviors that we will see repeatedly reenacted in the other major case studies this volume considers. Rumors ran riot in post-quake Lisbon. In the first few days, before order was restored by Portuguese troops, stories spread that "brigands" had descended like vultures upon the ruins of the city, including many escaped convicts from Lisbon's shattered prisons. These "mobs of ruffians and demons" were said to be committing wanton rapes, and in their greed they thought nothing of cutting off fingers to get rings and slicing earlobes to extract earrings from the dead and the living alike. While no one was accused of poisoning wells, Jews and *moriscos* were charged with deliberately setting fires in the ruins to flush out the remaining population and facilitate looting – or perhaps out of sheer malice and bloodlust.

There are also intriguing hints in some of the sources that anti-Protestant sentiment nearly spilled over into violence in post-quake Lisbon. Most Catholic Lisboners probably interpreted the quake as a judgment from God, and this sentiment was stoked by firebrand preachers, some of whom blamed the quake on both Lisbon's crass materialism and its toleration of foreign heretics. Penitent Catholics sporting crucifixes and sacred images filled the street, much to the dismay of the English expatriates of Lisbon, who feared that the "infatuated Mob" might try to "appease the Wrath of Heaven by our Destruction." After all, expatriate Englishman Thomas Chase wrote after the quake, "what was to prevent the crowd from suddenly turning on the Protestants … and exacting retribution for their misfortunes?" For this reason, Chase tried to keep off the ruined streets of Lisbon after the quake, as "it was impossible to guess what turn their furious zeal might make against the worst of criminals, a Heretic!"

Chase's caution seems to have been justified, as other English expatriates reported harrowing experiences at the hands of the "infuriated mob." One expatriate reported that, "when the crowds started to congregate around him and ask him questions in a language he did not understand," he feared that, as a Protestant, he would be "blamed for this catastrophe and flung to his death." Still another English merchant named Goddard was trapped by a "hysterical" Catholic crowd, which "surrounded and seized … [him] in the most violent Manner" and interrogated him in Portuguese. Goddard was afraid he was about to be lynched by the mob, but after a Catholic priest intervened, the Portuguese calmed down and even surprised Goddard with the sudden "tenderness of their behavior." It was only afterwards that Goddard realized what had probably happened – the priest had baptized Goddard as a Catholic! As Mark Molesky noted in his 2012 article on Goddard's encounter with the Lisbon mob, despite contemporary Protestant fears of Catholic persecution after the quake, there are no accounts from the period reporting any violent attacks on Protestants in Lisbon either on the day of the disaster or in the weeks and months to come. But the potential for such violence was definitely there.

Some outsiders did suffer violence in the aftermath of the Lisbon Earthquake, but it was inflicted, not by the mob, but by agents of the state. Shortly after the quake the Prime Minister, the Marquis of Pombal, sent the army into what was left of the city to restore order. It did so with brutal efficiency, hanging 34 looters in Lisbon's ruined streets as a deterrent against further criminality. Interestingly, only 11 of these men were Portuguese; the rest were foreigners or social outsiders, including 10 Castilians (mostly army deserters), 5 Irishmen, 3 Savoyards, 2 Frenchmen, a Pole, a Fleming (a Dutch speaker from Flanders), and a "Moor," who was probably a *Morisco* rather than an actual North African Muslim. In addition, a group of English sailors may have also been executed after being caught looting the ruins of a royal palace and chapel.

It is not entirely clear why foreigners and outcastes were disproportionally targeted for execution following the Lisbon Earthquake. Nicholas Shrady (2008) claims that foreigners were "easy scapegoats" in the quake's aftermath, a distraction from Lisbon's overwhelming problems. Making an example of foreign looters might have been a theatrical show of force designed to re-establish royal authority over the shattered town, or alternatively, it may have been an attempt to placate popular anti-foreign sentiment in Lisbon. It is also possible that some of the initiative came from the troops themselves. Presumably these soldiers were bombarded with disease cues in the burnt-out ruins of the corpse-filled city, so some evoked aversion to outsiders via the behavioral immune system was only to be expected. Given the desperate circumstances and the background of Catholic/Protestant tensions discussed above, in fact, it is surprising that reported incidents of violence against outsiders were so limited.

One reason for this, perhaps, was the swift and effective governmental intervention. Mind you, the earthquake seems to have inflicted lasting psychological trauma on the Portuguese King; although not in Lisbon at the time of the quake, Joseph I developed severe claustrophobia in its aftermath, to the point that he insisted on moving the royal court to an elaborate tent city he would later construct in the hills outside of Lisbon. However, the Marquis de Pombal stepped into this leadership vacuum and took total control of the relief and recovery measures, vowing to "bury the dead and feed the living." He froze grain prices, punished price gougers, distributed food, and ordered mass burials at sea. This last measure earned him the enmity of the Catholic establishment, which tried to insist on normal church burials despite the magnitude of the catastrophe. The Enlightenment-influenced Pombal also angered the clergy by downplaying the religious meaning of the earthquake, portraying it as an unforeseeable natural disaster rather than an act of divine retribution and pushing for urban rather than spiritual renewal in the quake's aftermath. One of Pombal's harshest critics, the charismatic preacher Gabriel Malagrida, proclaimed in a public pamphlet that

> It is scandalous to pretend the earthquake was just a natural event, for if that be true, there is no need to repent and try to avert the wrath of God, and not even the devil himself could invent a false idea more likely to lead us all to irreparable Ruin … it is necessary to devote all our strength and purpose to the task of repentance. Would to God we could see as much determination and fervor for this necessary exercise as are devoted to the erection of huts and new buildings!

But despite the clamors of the clerics, Pombal insisted on treating the quake as the work of natural forces rather than divine providence. Once the immediate crisis had settled down, in fact, Pombal sent out a comprehensive survey to all the parishes of Portugal inquiring as to when the quake struck, how long the shaking lasted, how much damage it caused, and other pertinent information. This exercise in information collection was not just a major breakthrough in the scientific study of earthquakes, but also has allowed modern seismologists and historians to reconstruct the Lisbon Earthquake in greater detail than any previous earthquake event. What is more, Pombal applied the same scientific principles to Lisbon's reconstruction. At Pombal's direction, the burned-out husk of medieval Lisbon was replaced by an entirely different urban landscape, featuring wide streets laid out on a grid and open-air plazas. As a defense against fire, the ground floor of all Baixa buildings were now required to be built of masonry. At the same time, Pombal insisted that Lisbon's new buildings be constructed around a sturdy internal wooden frame, which came to be known as a "Pombaline Cage." Pombal did not invent this architectural innovation – as discussed above, the *casa baraccata* dates back to at least the fourteenth century – but he did popularize it, even to the point of ordering Portuguese troops to march in formation around scale model buildings to test their resilience against quakes. Pombal was so successful in his plans to remake Lisbon that van Bavel et al. (2020) consider the Lisbon Earthquake as a case study example of Klein's (2008) "shock doctrine" – the tendency for elites to exploit the opportunities created by natural disaster to enact radical societal changes.

Pombal's reconstruction efforts notwithstanding, parts of Lisbon remained in ruins for decades after the quake. But that might have been for the best: according to Cynthia Kierner (2019), Lisbon's ruins were so attractive to sight-seeing northern Europeans that late-eighteenth-century Lisbon joined Rome and Paris as major destinations for young British travelers doing their "Grand Tour" of continental Europe. As a result, Lisbon may have a legitimate claim to be one of Europe's first "disaster tourism" destinations. Even today, visitors to Lisbon can walk under the remaining arches of Lisbon's Carmo Convent, which was partially destroyed by the 1755 earthquake but still stands, unused and roofless, in the heart of the Baxia district.

While Pombal was distracted for several years by the task of reconstructing Lisbon, he did not forget Malagrida, who was punished for his obstructionism and criticism three years later in an act of brutal revenge. Pombal accused Malagrida and some other troublesome priests of complicity with a failed assassination attempt on King Joseph I, who had been ambushed in his carriage by a masked assailant. Following four years imprisonment, Pombal arranged for Malagrida to be found guilty of heresy in 1761. Malagrida was strangled and his corpse was incinerated in a bonfire before his ashes were finally dumped into the Tagus. Since his true crime was opposing Pombal's vision of post-quake recovery, Malagrida could perhaps be seen as the last victim of the 1755 Lisbon Earthquake.

As for the quake itself, it would live on in the intellectual discourses of the European Enlightenment. At the time of the quake, many European thinkers had adopted Gottfried Wilhelm Leibnitz's doctrine of philosophical optimism, the idea that a beneficent God had created for us the best of all possible worlds. If evil exists in that world, Leibniz argued, it is not the work of God, though God may permit some misfortunes to occur if they contributed to the greater good. Some philosophers strained to fit the Lisbon Earthquake into this system of optimism; Jean-Jacques Rousseau, for example, thought that Lisboners had brought the disaster upon themselves by clustering together in "twenty thousand houses of six or seven stories," and even claimed that the earthquake had contributed to the general good by providing food for worms and wolves. Horrified by Rousseau's crass dismissal of Lisbon's suffering, the French *philosophe* Voltaire responded with his masterwork *Candide*, which demolished the optimism of Leibnitz (Voltaire, 1947). Far from being the best of all possible worlds, Voltaire claimed, the planet we share is dominated by human greed and misery and beset by forces outside of our control. The best we can do, Voltaire concludes in *Candide*, is to accept the world as it is and "cultivate our garden" – in other words, exercise what little control we do have to make our own lives better. Mark Molesky (2015) therefore argues that the Lisbon Earthquake and Voltaire combined to help shift the tone of the Enlightenment away from an earlier "Age of Reason" and toward a new "Age of Sentiment."

San Francisco, 1906

The Lisbon Earthquake of 1755, then, marked an important turning point in Europe, at least in terms of philosophy and architecture. In contrast, the San Francisco Earthquake of 1906 illustrates how easy it is to forget the past – and thus jeopardize the future – when we focus exclusively on the needs of the present. In a region where major quakes are a fact of life, San Francisco took few steps before the quake to combat seismic hazards. On the contrary, San Francisco's pioneer settlers magnified these inherent environmental risks by constructing hastily-built wooden houses atop sandy soils and fill. Worst yet, San Francisco learned few lessons from the 1906 earthquake. As a result, San Francisco remains acutely vulnerable to earthquake damage and may suffer an unprecedented catastrophe when the next "big one" inevitably arrives.

Building Vulnerability along San Francisco Bay

The city of San Francisco is the child of San Francisco Bay, a large tidal estuary on the coast of California and one of the best natural harbor sites on the U.S. West Coast. Ships based in San Francisco had access to the riches of the Orient, as the bay was only 5,000 nautical miles from Tokyo and 5,300 nautical miles from Shanghai. What is more, a natural navigable channel connected San Francisco Bay to the agricultural produce of the Napa Valley and California's Great Central Valley. The shore of the bay, therefore, was the perfect location for a major American city.

Perfect, that is, until you factor in the geological characteristics of the bay's swampy margins. As Doris Sloan (2006) points out, the first European settlers to San Francisco Peninsula found that the region consisted of little more than a few rocky hills surrounded by gravel slopes that overlooked a sea of dune sand. Closer to San Francisco Bay, the dunes gave way to tidal flats and marshlands, especially in the area around Mission Creek. As of 1847, this inhospitable terrain (then called Yerba Buena) had attracted less than 500 European settlers. However, everything changed in 1848. In that year, San Francisco, along with the rest of California, was incorporated into the United States in 1848 by the Treaty of Guadelupe Hidalgo at the end of the Mexican-American War. Almost simultaneously, gold was discovered in Sutton's Mill, and as many as 300,000 people flooded into California, hoping for quick riches in this land of opportunity.

While the gold fields were well inland from San Francisco, in the mountains beyond the great Central Valley, San Francisco Bay profited enormously from the gold strikes, as the city guarded the gateway to the California interior. By 1850, San Francisco's population had ballooned to 25,000, and just five years later crossed the 50,000 mark. Lacking solid ground for building construction, early San Francisco buildings were erected on wooden pilings or upon land raised above the tides with fill materials. The "South of Market" neighborhood of San Francisco, for example, was constructed by dumping garbage and nearly 200 feet of dune sand onto the streambed and surrounding marshes of the Mission Creek. Rapid growth and lack of urban planning also meant that early San Francisco's history was punctuated by frequent fires, including three great fires in 1850 that between them consumed most of the buildings of the city. Undeterred, San Francisco's pioneers gathered up the debris from the fires and used it to fill in the "water lot" coastal margins and reclaim even more valuable real estate from the Bay. As a rule buildings were both constructed and reconstructed as quickly and cheaply as possible, which meant using wood. By the end of the century, in fact, San Francisco had the highest percentage of wooden buildings (98%) of any city in the country. This might seem short-sighted, and it was: as Andrea Rees Davies (2012) points out, San Francisco was an "instant city" built by men who wanted to "transact the business of mining" and had little interest in making San Francisco their permanent home.

When San Francisco was struck by a major earthquake in 1868, the "instant city's" vulnerability to seismic shocks became abundantly clear. Even in the absence of an earthquake, the city's buildings were prone to subsidence: one newly-built theatre sank by two inches into San Francisco's fill soils in a matter of hours when it was weighed down by a large crowd on the night of its grand opening. These same soils liquified when impacted by the seismic waves of the 1868 quake, leading to numerous building failures, several serious fires, and half a dozen casualties. Damage was particularly bad in the "South of Market" district, where fill soil had been stacked atop already unstable alluvium.

A few contemporary observers blamed human greed for these unnecessary deaths. One San Francisco writer, for example, quipped that "men, in their eagerness to get to the front of the battle of the money-bags, have encroached upon the dominions of Neptune, until he has called his brother Pluto [God of the Earth] to his aid." Most civic leaders, however, conspired to minimize accounts of the damage and to blame it on easily rectified human construction errors, not natural causes or the inherent deficiencies of the landscape. All that was needed, they said, was to tweak the building designs. Masonry should be reinforced with iron rods, chimneys should be shorter, better brick and mortar should be used, more comprehensive fire codes should be established, and so on. But few such changes were implemented by San Francisco builders the years that followed, and even if such incremental improvements had been enacted, they would have done little to prevent catastrophe in 1906.

In any case, the 1868 earthquake was soon forgotten amidst San Francisco's headlong urban expansion in the latter decades of the 1800s. The Pacific Railroad reached San Francisco in 1869, accelerating the city's growth by giving it a direct railway connection to the Eastern seaboard. As a result, San Francisco's population reached nearly 300,000 by 1890, making it the nation's eighth-largest city, and would add another 100,000 or so by the time of the earthquake. Accommodating these new arrivals meant further expansion into unstable soils, especially the dune sand to the west of the urban core. It also necessitated building upwards – by the time of the earthquake, San Francisco's skyline was studded with hotels, tenements, and other wood-framed buildings four or five stories in height. Bizarrely, despite San Francisco's history of quakes and fires, very few of these high-rise buildings had sprinkler systems.

As San Francisco's population climbed toward 400,000, the population coalesced into neighborhoods with their own distinct identities, a process hastened along by San Francisco's own complex topography, which naturally cut the city into self-contained parcels. San Francisco's largest Italian neighborhood, dubbed the "Latin Quarter" in homage to Paris, lay between the Telegraph and Russian hills. The poor continued to live atop the unstable soils in low-lying South of Market, a neighborhood also known as "South of the Slot" after a cable car line called the "the slot" began to run up and down Market Street. South of the Slot was also an industrial zone of warehouses and factories, including a facility on Fremont Street where Levi Strauss manufactured his famous copper-riveted blue jeans. As always, wealth and privilege gravitated to the heights. A walker traveling from South of Slot up the slopes of Nob Hill would have been climbing both a topographical and a socioeconomic gradient, passing first through middle-class neighborhoods on the lower slopes before reaching the "faux-European monstrosities of the silver and railroad barons" (Fradkin, 2005) built along Nob Hill's fashionable summit.

The most famous of San Francisco's neighborhoods, however, was undoubtably Chinatown, which covered 17 city business and residential blocks in the heart of San Francisco. The Chinese began arriving in San Francisco in 1850, attracted by San Francisco's Gold Rush boom and the resulting economic opportunities. At first, they were welcomed to San Francisco, especially by railroad and mining companies who hoped to exploit the Chinese for cheap labor. By the 1870s, however, tensions began to grow between the Chinese and white San Franciscans, who blamed Chinese immigrants (with some justification) for the rise of prostitution, gang violence, gambling, and opium addiction. White workers also resented the Chinese for their willingness to work hard for relatively low wages, depressing San Francisco's labor market. These tensions boiled over into a race riot in 1877, the so-called "sand lot rally," which culminated in two nights of violence, four deaths, and over $100,000 in property damage. Although the riots were suppressed by city officials, resentment of Asians in San Francisco continued to fester and would be brought to a head by the 1906 earthquake.

The 1906 Disaster: Shake and Bake

Although few residents at the time were aware of the fact, San Francisco sits along an active fault line, the San Andreas plate boundary, where the massive Pacific Ocean Plate rubs up against the equally massive North American Plate to its east. As is typical of this type of transform fault, called slip-strike by scientists, the two plates had been locked in place for decades by friction, but tension had gradually built up within the plates themselves due to the cumulative motion of the mantle currents underneath. At 5:12am on April 18, 1906, the two plates finally slipped when potential energy stored within the deforming plates exceeded the force of friction, leading to sudden movement along 450 kilometers of the plate boundary. Estimates of the quake's moment magnitude range from 7.7 to 8.3; if the latter number is correct, it would mean that the San Francisco quake exceeded the maximum intensity quake that is normally generated along a transform fault. The quake unfolded over two main shocks followed by numerous aftershocks, and tremors were felt as far away as Oregon and Nevada.

As for people in San Francisco, who lived only 10 kilometers or less from the quake's epicenter, they felt much more than "tremors." Since the quake struck early in the morning, before many San Franciscans had even risen from their beds, most eyewitnesses to the quake were police officers walking their morning beats. One patrolman watched the undulating seismic waves ripple the streets and sidewalks as they approached: "it was as if the waves of the ocean were coming towards me, and billowing as they came." These first waves were probably the P-waves of the quake, since they did little damage to San Francisco's structures. Then the S-waves arrived and turned the world upside-down. "The ground seemed to twist under us like a top while it jerked this way and that, and up and down and every way," the same policeman reported. As in the 1868 quake, these seismic waves were felt especially strongly on unstable soils. Another police officer, Henry Powell, recorded a particularly striking description of the quake-time gymnastics of Valencia Street, a paved road perched precariously atop a layer of sand heaped upon the thick alluvium of old Mission Creek. "Valencia Street… began to dance and rear and roll in waves like a rough sea in a squall," Powell reported, but then "it sank in places and … vomited up its (trolley) car tracks and the tunnels that carried the cables. These lifted themselves out of the pavement, and bent and snapped. It was impossible for a man to stand," Powell concluded, "or to realize just where he was trying to keep standing."

Damage to buildings was little short of catastrophic. Powell watched in horror as the Valencia Street Hotel "lurched forward as if the foundations were dragged backwards under it" before "crumpl[ing] down over Valencia Street. It did not fall to pieces and spray itself all over the place," Powell continued, but rather the hotel "telescoped down on itself like a concertina [a type of accordion]."

Many other San Franciscans experienced these building collapses, not from outside, but from within. One survivor, William Stehr, was still in his room on the top floor of a five-floor boarding house South of the Slot when he felt a "bump, very sudden and severe" that brought down his ceiling. "The place fell in on top of me," Stehr remembered, "the breath seemed to be knocked out of my body and I went unconscious." When he eventually awoke, he found his feet were pinned beneath something and the "weight of the debris that covered my body was more than I could lift." With great effort, Stehr finally managed to free his feet. And not a moment too soon – screams of "fire!" were becoming audible through the rubble and debris, and Stehr caught the whiff of smoke in the air. Groping desperately, Stehr finally discovered a painful exit route from the ruins around him "through a sort of chimney that was bristling with nails and splinters."

Stehr emerged, bloodied by his struggles, into a world transformed. He had gone to sleep the night before on the fifth story of the hotel, but when he arose from its wreckage, Stehr found himself sitting "amidst the ruins nearly on a level with the street, and all around me was ruin and debris." But Stehr could count himself lucky. While the number of people killed by the initial collapse of San Francisco's buildings is not known for certain, since casualties in some districts like Chinatown were undercounted, it may have topped 3,000.

Despite the sudden severity of the quake, authorities sprang into action swiftly and decisively, though not very wisely. Believing that civic officials could not handle a disaster of this magnitude, the commander of the army division stationed in San Francisco's Presidio ordered his troops into the city to quell disorder and keep the peace. For its part, San Francisco civil officials ordered an immediate "shoot to kill" policy against looters, a decision that, like the army commander's decision to stage his troops into the city in peacetime, was later denounced as illegal and unconstitutional. In the days after the quake, trigger-happy soldiers, policemen, and militiamen opened fire against looters, suspected looters, and in some cases each other, with death toll estimates ranging from a dozen to a suspected high of 100 or more. Anecdotal evidence suggests that the poor and the Chinese were much more likely to be shot as looters than wealthy whites. In another sign of post-quake racism, anti-looting rules were not enforced in Chinatown, and since Chinese citizens were initially forbidden from returning to the neighborhood to protect their property, Chinatown was effectively picked clean by white "souvenir hunters" in the first few days after the disaster.

In the meantime, San Francisco began to burn. Fires sprang up everywhere after the quake, set off by "overturned stoves and oil lamps, damaged chimney flues, and broken electrical wires." One such conflagration, the devastating "Ham and Eggs" fire, was apparently started by a woman cooking breakfast on the morning of the quake. These opportunistic fires found abundant fuel in the ruins of San Francisco. Although some masonry and steel structures had been built since the fires of 1850, especially in the downtown financial district, 90% of San Francisco's buildings were still wooden in 1906. What is more, the two weeks before the earthquake had been almost rainless, so the timber-strewn ruins of San Francisco were tinder-dry at the time of the quake. As was the case in Lisbon, the combined blazes of San Francisco would eventually generate a firestorm, as rising columns of incendiary air created a relative vacuum that drew in ground level air and oxygen (Figure 2.3). According to author Jack London, an eyewitness to the inferno, even on an otherwise windless

Figure 2.3 San Francisco after the firestorm
Source: Still Picture Records LICON, Special Media Archives Services Division (NWCS-S), National Archives at College Park.

day, the "enormous suck" of San Francisco's fires pulled in air from the surrounding regions with the force of "half a gale."

Unlike many East Coast cities at the time, San Francisco did have a professional fire department, which boasted a dozen ladder companies and 42 fire engines, which in this era meant horse-drawn water pumps. But there was little water to pump, since the earthquake had damaged the pipes feeding water to the city's hydrant system. As a result, when fires began to rage out of control, the firefighters turned to explosives, in hopes of creating firebreaks that might contain the blazes. Luckily for San Francisco, plenty of explosives were available – this was a city founded on the mining industry, after all. Unluckily for San Francisco, firefighters were woefully untrained in the use of explosives. Worse yet, when dynamite began to run out, both firefighters and regular citizens turned to black powder explosives, which unfortunately generate heat upon ignition and thus were as likely to start new fires as to snuff old ones out. Ultimately, San Francisco's explosive firefighting tactics generated much sound and fury – witnesses would recall how the constant flashes of dynamite and rumbles of explosions during the San Francisco firestorm "gave the picture a suggestion of warfare" – but in terms of stopping the fires, they signified nothing.

By the time the fires were finally extinguished weeks after the quake, little of San Francisco was left. Flames had consumed 28,188 buildings in 514 city blocks which together covered 4.7 square miles. Fully half the city was gone, including most of the densely populated urban core, along with property worth somewhere between 500,000 million and 1 billion 1906 dollars. Damage was particularly severe to buildings like the Valencia Hotel that had been constructed atop the unstable soils South of the Slot, but once the fires began, they spread the destruction even to wealthier neighborhoods situated on higher and more solid ground. Many residents of San Francisco took advantage of the still-intact Pacific Railway to flee the ruins of the city after the quake, but as many as 250,000 homeless refugees remained, living in unsanitary conditions in over a hundred hastily built camps on the city's margins. These squalid camps added to the quake's death toll by incubating diseases: smallpox, typhoid (which may have killed 547 of the quake's survivors), and even *Y. pestis*, better known as bubonic plague (which killed another 77). Not surprisingly, given San Francisco's history of anti-Chinese agitation, San Franciscans placed the blame for these disease outbreaks squarely on the Chinese.

Aftermath: Racism and Boosterism in San Francisco

As is typical of major disasters, rumors ran riot in the aftermath of the quake. The popular press printed lurid accounts of inhuman "ghouls" descending upon the wreckage. Fradkin claims that "dusky-hued men," depicted as "grimacing Asians" or "a cross between African Americans and Italians" in contemporary illustrations, were accused of cutting off fingers to get at rings and even of "chewing diamond earrings from the ears of dead women." Thankfully, anti-foreign hysteria did not lead to mass killings in San Francisco, though as we saw above, the Chinese and other immigrant groups were probably targeted disproportionally by "shoot-to-kill" orders against looters.

Nonetheless, there were plenty of small-scale acts of violence, especially against the Japanese, who were beginning to replace the Chinese as the most feared immigrant group in the white San Franciscan mind. According to a report that the U.S. Commerce Secretary compiled for President Roosevelt, who at the time was worried about rising Japanese/American tensions, at least 19 Japanese immigrants were assaulted by white mobs in the aftermath of the 1906 quake, and this number did not include "minor attacks," including "stonings!" Even members of a visiting delegation of Japanese earthquake experts reported assaults; insults, dirt, and stones were repeatedly thrown in their direction, and one Japanese seismologist was attacked by a white assailant while on a fact-finding mission to Eureka. Most likely these assaults on the Japanese were just the tip of the iceberg in terms of anti-Asian sentiment, and many assaults on other groups went unreported because they did not raise the same foreign policy concerns for the Roosevelt administration as did American violence against the Japanese.

Nor did anti-Asian sentiment limit itself to the popular mob. When classes went back into session in 1906, the San Francisco School Board caved into pressure from parents and ordered that Asian students in San Francisco should be taught in separate schools, so that white children would not be negatively affected "by association with pupils of the Mongolian race." This same segregation was also applied to post-quake refugee camps. All Chinese quake survivors were herded into one single camp, which according to Jonna L. Dyl (2017) was "relocated multiple times under pressure from white property owners who did not want Chinese near their homes."

What is more, in the immediate aftermath of the quake, the city government made a shock doctrine style attempt to excise Chinatown from the San Francisco cityscape. In the white mind, Chinatown had long been a cancer within the flesh of San Francisco, and proposals to excise it had been floated as early as 1853. According to Dyl, many San Franciscans associated Chinatown with "filth, nuisances, and living conditions often compared to those of animals" – it was seen as an "alien urban ecology that posed a sanitary hazard for white San Franciscans." Small wonder, then, that city officials investigated the possibility of banishing Chinatown from the city's center and shifting it to the periphery in the weeks after the 1906 quake. One author, writing in support of this proposal, proclaimed that Chinatown's destruction was in fact an act of "divine wisdom," celebrating the fact that "fire has reclaimed to civilization and cleanliness the

Chinese ghetto." In the end, however, Chinese elites in San Francisco and their white allies successfully defended Chinatown from removal plans.

When San Francisco was rebuilt, therefore, it still included Chinatown. What it did not include was any significant improvements in terms of seismic protection. As was the case after the 1850 fire and the 1868 earthquake, rebuilding quickly took priority over rebuilding better. In the words of civil engineer John Galloway, a member of the citizens committee overseeing the rebuilding of San Francisco, "What San Francisco needs is the cheapest building possible in which business can be done, to ensure that the community has enough to eat. The other subjects can wait." Instead of addressing the issue of unstable fill soils, for example, San Francisco builders actually worsened the problem by using debris from the 1906 quake to fill in yet more coastal regions, including the Marina district in northern San Francisco near Fisherman's Wharf. In addition, while new building codes calling for some lateral reinforcement of structures were enacted in 1906, most builders ignored them, and in any case the new standards were quickly watered down in subsequent building code revisions.

As for the threats posed by future earthquakes, they not only went unaddressed, they were not even acknowledged. Just six days after the earthquake, the powerful San Francisco Real Estate Board met and decreed that "the calamity should be spoken of as 'the great fire,' and not as 'the great earthquake'." This proclamation soon rose to the status of orthodoxy in San Francisco, advocated by newspapers and powerful institutions such as the Pacific Railway. For Christian Pfister (2011), the roots of this attitude were psychological, the desire to shift the roots of the disaster away from uncontrollable nature and into the realm of controllable human causes.

Well, perhaps. More likely, these gathered real estate agents had a quite different motive: in order to protect their own bottom line, they wanted to normalize the disaster, to present it as a random calamity that could happen at any time in any U.S. city rather than the result of a predictable hazard endemic to the U.S. West Coast. A climate of boosterism, therefore, helped to prevent San Francisco from creating a "culture of disasters" around earthquakes. The result was a lackadaisical attitude toward seismic threats "not shared by any other earthquake district in the world," at least in the estimation of United States Geological Service scientist Grove Karl Gilbert (Fradkin, 2005). Were such a culture be allowed to develop, Gilbert noted, then California might gain a reputation for seismic instability, and as a consequence "the flow of immigration will be checked, capital will go elsewhere, and business activity will be impaired." San Francisco's culture of disaster, therefore, was smothered in its cradle by a culture of capitalism.

Great Kantō, 1923

Just 17 years after the San Francisco quake and firestorm, an eerily similar disaster played out in Japan's Kantō Plain on the other side of the fast-moving Pacific Ocean Plate. As in San Francisco, the Great Kantō Earthquake struck a region experiencing massive and unplanned urban expansion. What is more, in both events, the quake's seismic waves struck timber-built cities constructed on unstable soils, setting the stage for post-quake firestorms. But there were differences as well, which illustrate the important roles that economic, demographic, political, and above all cultural factors play in shaping the outcome of a major natural disaster. Demographic factors, for example, ensured a much higher death toll in Tokyo and Yokohama, both net and per capita, compared to San Francisco. In addition, while there was certainly a rise of anti-immigrant tension following the San Francisco quake, political and cultural factors in Japan, combined with the human tendency toward xenophobia in the presence of disease cues, would trigger a full-blown anti-foreigner massacre.

Building Vulnerability on Tokyo Bay

When American Admiral Commodore Perry sailed his "black ships" into Tokyo Bay in 1852 in his mission to open Japan to western trade, he encountered a society that was essentially medieval in social and economic structures. Since 1603, Japan had been under the control of the Tokugawa Shogunate, which had preserved internal peace (and the domination of the landholding *daimyō* and samurai elites) in part by closing Japan off from outside influences and sharply restricting the scope of Japan's international trade. Within two decades of Perry's visit, however, the Tokugawa shogunate collapsed, overcome by internal rebellion and foreign pressure. Its replacement, the regime of Emperor Meiji, called itself a "restoration." However, its goals were nothing short of revolutionary: reformers in the imperial court pushed for the rapid transformation of Japan into a centralized, westernized state capable of fending off the technologically superior western powers. By and large, they succeeded. By the 1920s, Japan could boast of a thriving industrial sector, over 12,000 km of train lines, parliamentary institutions, and a modernized army and navy which had defeated a major European power, Russia, in a 1904–1905 war.

Japan's rapid development also transformed the cities of the Kantō Plain, the area most impacted by the 1923 earthquake. At the time of Perry's visit, the Japanese capital of Edo was already one of the world's largest cities at 1 million inhabitants, but by the early twentieth century, Edo (renamed Tokyo) had bloomed to over 2 million Japanese, drawn there by the economic opportunities created by administrative centralization and rapid industrialization. Yokohama's growth was even more dramatic. In Perry's day, the land that would become Yokohama was little more than "a marsh by the edge of a deserted bay," in the words of an early British envoy to Japan. The Meiji needed a new port for Japan's increased foreign trade, however, and Yokohama's location seemed ideal, since it was close enough to serve the economic needs of the capital but also kept foreigners at a safe distance from Japan's strategically important Tōkaidō Road between Tokyo and points north. By the 1920s, Yokohama's population had rocketed to 423,000, in part, because the Japanese silk exports boomed in the late 1850s after Europe lost most of their own silkworms to epidemic disease.

While most of the burgeoning population in both cities was Japanese, the Kantō Plain did host sizable expatriate populations as well, especially Yokohama, Japan's most cosmopolitan city. In 1923, Joshua Hammer (2006) notes, a walker on the streets of Yokohama might have rubbed shoulders with "Bengali shopkeepers, Lebanese Jewelers, Chinese money-changers, Russian optometrists, French diplomats … English silk merchants, Spanish missionaries, [or] American tea traders" – and of course western newspaper correspondents like W. Reer Harris of the Kokusai News Service. Yokohama also hosted a colorful group of aristocratic Russian exiles who had fled the Bolshevik Revolution and were now forced to earn their keep as "barmaids, dance instructors, or piano teachers."

However, Koreans were probably the largest group of expatriate workers in Japan at the time of the 1923 quake. They were also the most despised. After defeating both Chinese Imperial troops and indigenous Korean rebels with its modernized armed forces, Japan had formally annexed nearby Korea in 1910. Koreans were now in theory full Japanese citizens, with the right to migrate to Japan for work, and many of them did just that during the labor shortage of the late 1910s, when Japanese industries actively recruited Korean workers. By 1920, roughly 200,000 of these mostly-male Korean expatriates were working in Japan, 12,000 of whom lived in the cities of Tokyo and Yokohama.

Overall, the situation of Koreans in Japan was not unlike that of the Chinese in San Francisco. Economic and social discrimination ensured that Koreans performed low-skill manual labor and were concentrated in poor neighborhoods, dubbed "Koreatowns." Their willingness to accept low wages made them popular with employers but hated by Japanese workers, who saw them as an economic threat, especially after cycles of economic depression set in during the early 1920s. What is more, since 1919, Japanese rule over Korea had been rocked by revolts and assassination plots, including an attempt to kill the Japanese governor of Korea during a military parade in Seoul. Rumor had it that the Korean guest workers in Japan were hatching similar conspiracies. According to Hammer, a combination of economic competition and political tensions meant that Japanese attitudes towards Koreans had "congealed into a toxic mix of guilt, fear, and resentment" by the time the earthquake struck.

The Japanese, Koreans, and other residents of both Yokohama and Tokyo occupied an urban landscape that blended traditional and modern architectural elements. Both Tokyo and Yokohama hosted a large number of factories, warehouses, and European-style multistory masonry structures, a few of which (like Frank Lloyd Wright's Imperial Hotel in Tokyo) were designed with seismic protection in mind. Yokohama's coastline was also studded with wharves and dockyards, while central Tokyo boasted embassies, government ministries, and the imposing Imperial Palace and the halls of the Imperial Diet.

However, such modern edifaces were the exception, not the rule. About 90% of structures in both cities were homes built according to traditional Japanese architectural principles, which meant timber frames, ceramic roof tiles, and oiled paper paneling. This sort of construction reflects Japan's long experience with the region's natural hazards: the use of light materials, ample lateral bracing, and a low profile rendered such houses resistant to both typhoon (cyclone) winds and to seismic shaking. However, houses of this sort were highly vulnerable to fire. This may not have been much of a problem in traditional Japanese villages, where sufficient space could be left between houses as a fire break, but it was a recipe for disaster in densely-packed urban areas like Tokyo and Yokohama. Urban fires had already become such frequent occurrences in Tokyo before 1923 that they had been euphemistically dubbed "flowers of Edo" by the Japanese. Combatting these frequent fires was the job of public and private fire brigades which, like the rest of Japan, had embraced western-style equipment and techniques, including portable pumps and a hydrant system.

As in San Francisco, the vulnerability of Tokyo and Yokohama's urban landscape to natural hazards was compounded by weak underlying soils. Tokyo consists of two broad zones, the *Shitamachi* or "low city" on the river delta to the east of the Imperial Place and the *Yamanote* or "high city" in the hills to the palace's west. While much of the *Yamanote* rests upon solid rock, the *Shitamachi* is a marshy zone of alluvial soil and landfill that is highly susceptible to earthquake shaking. As in San Francisco, the distinction between low city and high city was as much a matter of social class as topography: the *Yamanote* was traditionally dominated by the landed elites and their retainers while the *Shitamachi* was mainly inhabited by lower-status merchants and craftsmen. The underling soils were even worse in Yokohama, which was built upon marshland, rice paddies, and coastal landfill.

September 1, 1923

Unfortunately for the inhabitants of Tokyo and Yokohama, these unstable soils lie dangerously close to an active seismic hotspot, the four-way juncture of the Philippine and Pacific Sea oceanic plates, the Eurasian continental plate, and the relatively small Okhotsk plate. While the Eurasian and Okhotsk plates are in no particular hurry, the Pacific plate scurries along at a blazingly-fast 10 centimeters a year, diving as it does beneath the Okhotsk and Philippine Sea plates. This sort of plate interaction is the worst-case scenario for nearby populations: not only do subduction zones generate the strongest earthquakes, the high relative speed of the interacting plates ensures that such earthquakes are fairly frequent occurrences, on geological if not human time scales. As we will see in Chapter 3, this same plate boundary would generate an even stronger quake in 2011, triggering the Tōhoku tsunami.

The Great Kantō Earthquake did trigger a deadly tsunami of its own, which devasted Japanese communities along Sagami Bay with wave crests that might have reached 13 meters in height. Luckily for Tokyo and Yokohama, the Miura Peninsula prevented most tsunami waves from entering Tokyo Bay proper. The same could not be said, however, for the seismic waves of this magnitude 7.9 quake, which inflicted serious damage throughout the Kantō Plain. One notable witness to the severity of the quake was Japanese seismologist Akitsune Imamura, who had written in 1905 that a major quake beneath Sagami Bay was overdue and that it might kill as many as 100,000 or even 200,000 people in overcrowded and ill-prepared Tokyo and Yokohama. Eighteen years later, Imamura experienced the very event he had predicted and survived to say "I told you so." In his later recollections about the earthquake, Imamura wrote that it began with "slow and feeble vibrations," but then escalated rapidly, shaking the sturdy office building around him "to an extraordinary extent." After the 12th second of the quake, the vibrations become "slower but bigger," before eventually declining to the point that they resembled an "undulatory movement like that … on a boat in windy weather."

While the Tokyo building Imamura occupied apparently survived the initial quake, many others did not, especially in Yokohama, where the underlying soil was particularly prone to liquification. The best account of the impact of the quake in Yokohama comes from the pen of American businessman Otis Manchester Poole, who was in the offices of the Dodwell & Co. trading house when the earthquake struck. From the relative safety of an interior doorway, Poole watched as "slabs of plaster left the ceilings and fell about our ears … walls bulged, spread, and sagged; pictures danced on their wires, flew out and crashed in splinters," while furniture such as desks, cabinets and even heavy safes "spun for a moment and fell on their sides. It felt," Poole remembered, "as if the floor were rising and falling beneath one's feet in billows knee high." But worse was to come: moments later, "as if heralding the end of the world, the earth seemed to rise in the air and rock; and all around us thundered the deafening roar of cascading buildings."

When the shaking finally subsided, Poole and his colleagues emerged from their office building into a Yokohama they no longer recognized. The sight that greeted him when he reached the street was like a "blow to the stomach" – main street was now an unrecognizable tangle of fallen beams, cracked masonry, roof tiles, and other detritus. But "what seemed most terrible" to Poole was the quiet, a "deathly stillness ... into which the scraping of our own feet sounded ghostly. Shattered fragments of buildings rose like distorted monuments from a sea of devastation," over which had settled a thick blanket of dust, "and through the yellow fog of dust, still in the air, a copper-colored sun shone upon this silent havoc in sickly unreality." Overall, about 12.4% of Yokohama's buildings collapsed during the initial quake. Damage was particularly bad in Yokohama's coastal Kangai neighborhood, which had been constructed atop a filled-in shallow inlet near the center of Yokohama. By the time of the quake, Kangai was packed tight with the flimsy wooden dwellings of poor Japanese and expatriate laborers who, in the words of Hammer (2006), had been lured to Yokohama "by the boundless economic opportunities of the Far East's greatest port." As for Tokyo, which was farther from the quake epicenter and partially built upon harder rock, the percentage of buildings toppled by the initial tremors was much lower, perhaps only 1%. In general, Japanese-style buildings constructed of wood weathered the quake much better than masonry buildings. Unreinforced concrete structures fared worst of all: a foreign correspondent for the *Atlantic Monthly*, for example, watched in horror as through the windows of his office as a "four story concrete building vanished, disintegrated in the flash of an eye."

No sooner had the quakes subsided than the fires began. In part, this reflects the bad timing of the quake, which struck just before noon, just as lunch was being prepared on traditional charcoal braziers in both cities. As Edward Seidensticker (1983) points out, however, many of the fires had more modern causes, such as chemical explosions, ruptured gas cannisters, or electrical discharges from downed power lines. Within 30 minutes of the quake, at least 136 fires were burning in Tokyo alone. In both cities, especially Yokohama, these fires were stoked and spread by the gusting winds of an offshore typhoon, the edge of which brushed against the Kantō Plain region while it passed north of Japan. Worse yet, as Imamura had predicted back in 1905, the earthquake broke the water mains in both Tokyo and Yokohama, rendering the fire hydrant system inoperable. Even if water had been available, however, it is unlikely that firefighters could have done much to save either city, given the sheer scale of the developing inferno.

In contrast with the San Francisco earthquake, where building collapses accounted for most of the casualties, raging fires in Tokyo and Yokohama killed many more people than the quake itself. Flames would ultimately consume 50% of Tokyo and fully 80% of Yokohama, and the cumulative loss of property and buildings was estimated as equal to fully 5% of Japan's annual GDP. In addition, fire was responsible for most of the 142,000 lives claimed by the disaster – Imamura's estimate that the next big quake would claim 100,000–200,000 casualties, therefore, proved to be chillingly accurate.

This high mortality rate raises an interesting question: why were the Tokyo and Yokohama firestorms so deadly, while the firestorm in San

Francisco claimed relatively few lives? Part of the explanation is demographic: Tokyo and Yokohama combined hosted nearly six times the population of San Francisco, so more casualties were only to be expected. What is more, the area engulfed by flames in Tokyo (34.7 km^2) was much larger than in San Francisco (10 km^2), making escape from the inferno more difficult. Someone in the center of burning region of San Francisco, for example, would have had to travel about 1.8 kilometers to reach the edge of the burning district, while the same person would have had to travel closer to 3.3 kilometers to escape the flames in Tokyo. In addition, it was easier to stay ahead of the fire in San Francisco, where streets were fairly wide and generally arranged in a grid. While this was true of some parts of Yokohama and Tokyo as well, both cities also included neighborhoods that were described as "warren-like" by European observers, with narrow streets and severe overcrowding. Making matters still worse, Tokyo's "low city" was divided up by rivers and canals that could only be crossed at bridges, but many of these bridges collapsed or caught fire during the disaster, blocking escape.

Tokyo did have some parks and other open areas within the city that should have provided safe havens from the firestorm. Unfortunately, one such open space became the scene of the disaster's deadliest tragedy. As the flames consumed Tokyo's low city, police herded fleeing refugees to the grounds of the recently dismantled army clothing depot, an open space in an otherwise crowded city. By mid-afternoon, as many as 45,000 people had crammed themselves and their belongings into the depot grounds, which seemed to be safely upwind of the spreading fires. By 3pm, one witness later remembered, "the wide space was so packed with people that they formed a solid block, within which individual movement became impossible." Unfortunately, around the same time, the prevailing wind began to change direction, possibly due to the movement of the offshore typhoon. Soon, an "unbroken circle of fire" surrounded the depot grounds, making escape impossible.

What the trapped crowd did not know, but was about to find out, was that an unusual atmospheric phenomenon was forming in the firestorm convection cell over Tokyo. Tornados require two elements: an updrafting column of air, and an overhead layer of cumulous clouds. Thanks to the heat of the fires and the thick, billowing smoke clouds that now hung above Tokyo, both elements were now in place. The fire tornado or "dragon twist" that thus formed over Tokyo was up to 200 meters tall, the width of a sports stadium, and spun fast enough to send barges on the Sumida River flying into the air. And at about 4pm, to the horror of the crowd, this dragon twist descended upon the depot grounds. Only about 300 of the refugees in the army depot would survive; the rest were either battered to death, suffocated, or burned alive by the twist's superheated, oxygen-starved gales.

When the dragon twist withdrew, it left behind a scene of almost unbelievable carnage. According to American reporter Henry W. Kinney, bodies in the depot were "twisted and contorted, naked or with only rags clinging to them, covering acre upon acre." Since there was not enough room on the ground for all the dead, some remained standing in place, "the dead rubbing elbows with the dead." In the subtropical late summer sun, decomposition quickly set in, and with it legions of flies and millions of maggots. In the end, rescuers gave up on burying the dead individually and burned the corpses where they lay, finishing the job that the dragon twist had started.

Post-Quake Massacre and Mitigation

In the aftermath of the quake and fires, rumors ran riot among the survivors. The quake, some claimed, was a punishment for Japan's growing materialism or else a divine judgment against Japan's recent westernization. Mount Fuji was about to erupt, others claimed, and a tsunami was imminent. Still other rumors spread that the Koreans were raping women, poisoning wells, starting fires, planting bombs, or even planning a general insurrection against the Japanese government's shaken authority. As Nihnee Lee (2008) points out, these rumors about the Koreans struck a chord and snowballed in a way that the others did not, provoking mass panic and retaliatory violence against the hated Koreans. As a result, the embers of Yokohama and Tokyo were still warm when a new phase of the tragedy began to unfold, the murder and massacre of much of the Kantō Plain's Korean population.

The Korean massacre began just one day after the earthquake. That morning, the streets of both Tokyo and Yokohama were abuzz with the rumor that Korean convicts had broken out Negishi prison on the outskirts of Tokyo and were marauding through city, raping women and poisoning wells as they went. The poisoning rumors may have seemed plausible to many Japanese that day as much of the well water in Tokyo and Yokohama had become cloudy with sediment, a commonly-seen phenomenon following the vibrations of a major quake. Outraged Japanese armed themselves with bamboo spears, kitchen knives, hammers, and wooden clubs and formed vigilante mobs to hunt down Koreans, who were identified by their accents, their fuller beards, or their inability to recite the list of previous Japanese Emperors (Figure 2.4). Koreans who fell into the hands of the mob were subjected to unimaginable violence. According to Sonia Ryang (2003):

> From September 2 onwards, bodies began to appear in the street corners and rivers, bodies that presented with different causes of death from those killed by the earthquake and the fire that followed. They were obviously murdered bodies, as their hands were tied at the back and vivid marks of torture covered every limb, every piece of flesh, marked by broken bones, the belly cut open, eyeballs gouged out, and body parts severed and gone.

Figure 2.4 Vigilante activity after the Great Kantō Earthquake, 1923
Source: Chronicle of World History/Alamy Stock Photo.

As for the government, its role in the massacre was somewhat contradictory. In the immediate aftermath of the quake, the government did order newspapers to print announcements threatening punishment against those spreading "untrue stories" such as Korean uprisings and atrocities. At the same time, local police played an important role in spreading anti-Korean rumors and may have also played a part in the massacres, for example, by helping to organize "defense squads" that acted more like lynch mobs, or by rounding up Koreans for their "protection" and then allowing them to fall into the hands of the bloodthirsty mob. One of the most outrageous massacres actually occurred within the confines of a police station on September 4, when men armed with "swords, carpenter hooks, and spears" tortured and killed about 100 unresisting Koreans – men, women, and children – in the presence of police. Army soldiers may have played an even more active role in the massacres. Many troops sent into Tokyo and the Yokohama area after martial law was declared on the 2nd of September treated it as a military operation rather than a rescue mission, seeking out and executing Korean "national traitors" in the smoldering ruins. However, the Japanese army did eventually play a crucial role in stopping the massacres by distributing flyers denying the anti-Korean rumors and (eventually) by actively subduing rioters.

Although the Japanese government would later try to claim the Korean massacre victims numbered only in the hundreds, historians have put the actual death toll at 6,000 or above, or somewhere between a third and a half of the pre-quake Korean population of Tokyo and Yokohama. About 700 Chinese residents of these cities were slaughtered as well, as were some Japanese who were mistaken for Koreans, including two deaf teenagers. Army and police officials also took advantage of the chaos to abduct and kill a number of Japanese leftists and labor organizers, who were accused of colluding with the traitorous Koreans.

So why did the Japanese follow up their devastating natural disaster with an equally tragic man-made encore? Scholars have proposed a number of theories, though none are fully convincing. Noting that the massacre specifically targeted Koreans rather than other migrant groups present in Japan, historian J. Michael Allen (1996) has claimed that the massacre was the result of a hatred born from Japanese colonial occupation of the Korean peninsula. This seems sensible enough, though it does not fully explain why the massacre coincided with a natural disaster. Anthropologist Sonia Ryang, on the other hand, has depicted the Korean massacre as a manifestation of a specifically Japanese variation of nationalism which still has salience even in modern Japan. Drawing on Giorgio Agamben's concept of the *homo sacer*, a term meaning cursed and polluted beings who are outside of the normal human social order, Ryang argued that Koreans, at the time of the earthquake, were understood as "outcaste" non-persons whose deaths reinforced Japanese sovereignty. For Ryang (2003), the status of Koreans as polluted outcasts explains not only why so few rapes occurred during the post-quake massacres, but also why the Koreans were subjected to both murder and brutal torture. In contrast, Japanese leftists, who were national enemies but not dehumanized outcastes, were merely executed.

Ryang's argument is valuable insofar as it provides insight into the cultural lens through which Koreans in Japan were perceived. However, her claim that the Korean massacre was an idiosyncratic product of distinctly Japanese cultural traits seems unnecessarily complicated. Her theory also overlooks the fact that the Kantō Massacre is just one of many cross-cultural examples of xenophobia-fueled violence occurring in the wake of disasters and disease. I am not arguing culture played no role in this massacre – quite the contrary, cultural factors were instrumental in the outcome as they directed the attention of the crowd toward the Koreans by assigning them the status of outsiders. However, I would contend that the human behavioral immune system offers the most straightforward explanation for the crowd's violent post-quake xenophobia. Disease cues, after all, were omnipresent in the aftermath of the Great Kantō Earthquake and inferno, and to a far greater extent than was the case in San Francisco in 1906. Death, blood, and decomposition were everywhere, and even the living were marked with the dark bruises of injury and the scorch marks of fire. In addition to explaining the scale and the timing of the massacre, the BIS theory helps us understand why well-poisoning rumors seemed so credible: the behavioral immune system sensitizes people to food purity and contamination threats. Finally, the BIS theory offers an explanation for the phenomenon noted by Ryang, the fact that few rapes were reported during the killings. The main emotion triggered by the BIS is disgust, which might predispose someone to murder, but not to rape.

Given the atrocious violence of the Korean Massacre of 1923, it is small wonder that many modern Japanese politicians now dispute the scale of the event or even deny it ever happened at all. Certainly, it is nothing to be proud of. However, the Japanese can look with pride at another legacy of the quake: unlike the San Francisco quake of 1906, the Great Kantō Earthquake of 1906 helped to create a robust culture of disaster designed to mitigate or even prevent future earthquake and urban fire disasters. As protection against fire, entire neighborhoods in Japanese cities must by law be constructed of fireproofed materials like reinforced concrete, in hopes that during major conflagration these areas will serve as firebreaks. In a similar vein, Japanese urban planners have designed and embedded green spaces into cities to serve both as firebreaks and evacuation points. In addition, modern Japanese cities are dotted with "disaster-prevention" bases, well-stocked facilities strategically located near possible disaster zones from which to launch disaster relief and recovery operations. The same facilities also have an educational mission, providing training to the local populations and serving as centers for the dissemination of information about Japan's natural hazards. These same "disaster-prevention" bases also play a crucial role in Japanese tsunami preparedness measures, as we will see in Chapter 3.

Perhaps most importantly, in the period since 1923, Japan has been a pioneer in the field of earthquake-resistant construction. Most new Japanese buildings feature a steel internal frame, which is more flexible against earthquake shocks than reinforced concrete, as well as diagonal dampers to mitigate lateral shaking. Many new buildings feature an enormous pendulum in the core or near the roof of the building, designed to counter seismic waves by canceling out earthquake oscillations, and some older buildings have been retrofitted with pendulum systems as well. Buildings in the most seismically active areas of Japan are constructed upon layers of lead, steel, and rubber, allowing the base of the building to move independently of the earth below and thus buffer the building itself from tectonic forces. Some Japanese buildings even rest upon silicone oil-filled pistons, which absorb kinetic energy during an earthquake and transform it harmlessly into heat.

So why did the Kantō Plain Earthquake of 1923 help to create a "culture of disaster," while San Francisco's 1906 earthquake did not? As Grove Karl Gilbert noted in the last section, San Francisco's strong culture of boosterism discouraged serious post-quake reflection, which might have been bad for business. Different historical experiences probably played a role as well. The 1923 disaster in Japan occurred in a society that already had already developed a culture of disaster because of considerable previous experience with quakes and fires, and thus was not starting from scratch. In contrast, most San Francisco residents were recent arrivals from the relatively quake-free East, who were probably predisposed to interpret the 1906 quake as an anomaly or "act of God" rather than a recurrent threat. The much greater scale of the Japanese disaster also may explain why Japan was more willing than San Francisco to enact changes in their quake's aftermath. The high estimate for the San Francisco quake, 3,000 deaths, pales in comparison with the 142,000 who died in the Great Kantō Earthquake, and even if the deaths are presented as a fraction of the whole urban population, the Great Kantō Earthquake (.53%) was overwhelmingly more deadly per capita than the San Francisco earthquake (.075%).

While these factors no doubt played a role, I would speculate that Japan's more robust culture of disaster also reflects Japanese/U.S. differences in socioecological psychology. Japan and the United States sit on opposite ends of the collectivist/individualist spectrum, which may help explain why Japan was more willing to mandate post-disaster policies and regulations that prioritized the public good. In more individualist San Francisco, mandates of this sort, however well intentioned, might have been seen as unacceptable violations of personal rights. Japan is also a much "tighter" society than the United States in general, and California in particular, meaning they are more tolerant of rules (such as zoning laws and construction standards) and stricter about enforcement. It may also be that Japan's more robust disaster preparations reflects Japan's tendency toward slow life strategies (high investment in fewer offspring, pro-social behavior, future orientation, etc.), in comparison with faster California. While good data on national tendencies in life history strategies are lacking, research does suggest that slow life strategies strongly correlate with population density – and with an estimated 1,300 people per square kilometer, Japan's Kantō Plain is one of the most densely packed places on Earth. The comparable figure for New Jersey, the most densely populated state in the U.S., is only 488 people per km^2.

Haiti, 2010

The earthquake that rocked Haiti's capital in January of 2010 was, in the estimation of economist Andrew Powell, "the most devastating catastrophe that a country has experienced possibly ever" (Donlon, 2012), at least when the death and damage were considered proportionally to pre-quake population and GDP. While the Kantō Plain Earthquake might have inflicted damage to Japan equal to about 33% of its pre-quake gross domestic product, the overall damage done by the 2010 Haitian quake amounted to 117% of Haiti's GDP. Yet, despite the severity of the damage done to Haitian people and property, the 2010 Haitian earthquake was by far the weakest of the four case study quakes considered in this chapter.

The Haitian earthquake reached a moment magnitude of 7.0, meaning it was only about one-eighth the strength of the 7.9 Kantō Plain quake.

So why was this smaller quake proportionally more destructive? The answer lies within the familiar equation of Disaster = Hazard + Vulnerability. The waves of the 2010 Haitian quake shook a society that was extraordinarily vulnerable to seismic hazards due to a combination of historical inequities, governmental corruption, economic marginalization, environmental degeneration, and architectural errors. Taken together, these elements of vulnerability combined to create the second worst natural disaster of the twenty-first century, behind only the 2004 Indian Ocean Tsunami.

Haiti's 500-Year Earthquake

Many of the roots of Haiti's 2010 disaster reach all the way back to the colonial era. Haiti started its existence as Saint-Domingue, a colony created when Spain ceded the western end of the Caribbean island of Hispaniola to France in 1625. Once in charge, French planters imported hundreds of thousands of African slaves to grow sugar and coffee in Saint-Domingue's subtropical soils. By the 1780s, thanks to the labor of almost half a million enslaved Africans, Saint-Domingue had become France's most valuable colony, producing 40% of the sugar consumed in Europe and fully 60% of Europe's coffee. Nonetheless, little of that wealth remained in Saint-Domingue. While some revenue went into the pockets of Saint-Domingue's white and free African population, much of the rest was siphoned off by the French state or non-resident landowners. In any case, Haiti's wealth came at a heavy cost, both human and environmental. Thanks to Saint-Domingue's brutal labor regime and resident diseases, Haitian slaves suffered a 6–10% annual mortality rate, the worst such rate in the Western Hemisphere. What is more, sugar production is extremely energy-intensive, so Saint-Domingue's native forests began to disappear, sacrificed to the insatiable appetite of the sugarhouse boilers.

French control over Saint-Domingue and slavery were toppled during the chaotic years of the French Revolution. By 1791, Haiti's slaves were up in arms, led in part by General Toussaint Louverture. Although the powerful Louverture pledged allegiance to France, the French under Napoleon would send a huge expeditionary force in 1802 to re-take Saint Domingue and restore white rule (as well as its former profitability for the French state). The French did succeed in capturing Louverture, but their troops were decimated by Caribbean diseases and a successful guerilla campaign, forcing their withdrawal in late 1803. As they left, General Jean-Jacques Dessalines declared himself emperor for life and re-named his realm "Haiti," an old Indigenous Taíno word meaning "land of the mountains." He inaugurated his rule by ordering a genocide of most of the few whites that remained in Haiti. In the meantime, like Louverture before him, Dessalines tried to re-establish the planation system in Haiti; while wages were paid, Haiti's former slaves were attached to specific plantations without freedom of movement.

Although the French had withdrawn, they had by no means conceded Haiti's independence, and in this they were joined by the British and the United States, who still practiced slavery in their territories and thus saw Haiti's very existence as a free nation as an existential threat. As a result, France, Britain, and the United States imposed a trade embargo on Haiti in an attempt to strangle the infant nation. This embargo was only lifted in 1825 after Haiti agreed to pay exorbitant financial reparations to compensate France for its lost human property. Haiti's debt, which was not fully paid off until 1947, crippled the young country; in 1900, for example, an astounding 80% of Haiti's government expenditure went toward servicing Haiti's debt. As a result of these neocolonial exactions, Anthony Oliver-Smith (2012) claimed, Haiti "went from being the richest Caribbean colony, 'the pearl of the Antilles,' to the most impoverished nation in the Western Hemisphere."

Haiti's history between 1825 and the 2010 earthquake reads like a catalogue of woes. Chronic instability in Haiti, and fear of European adventurism, provoked Woodrow Wilson to send the marines into Haiti, which was occupied by the United Sates from 1915–1934. American occupation brought some benefits, such as improved roads and transportation infrastructure, but these were built using forced labor, provoking considerable Haitian unrest. The massive growth of Haiti's capital of Port-au-Prince also dates to this period, a byproduct of American centralization efforts. American withdrawal after 1934 led to renewed instability and a succession of strong-man governments, culminating in the dictatorship of "Papa Doc" Duvalier from 1957–1971, then his son, known as "Baby Doc," from 1971–1986. Papa Doc's reign was brutal even by Haitian standards: he relied on Haitian Vodou and a paramilitary force called the *Tonton Macoutes* to cow his political opponents while lining his own pockets with state funds. Under Baby Doc's regime, which Oliver-Smith has dubbed a "kleptocracy," repression was relaxed but public corruption reached a new apex. Patrick Bellegarde-Smith (2011) has dubbed Haiti's economy "*un capitalisme sauvage*, a rudimentary form of capitalism in which majorities starve and the rich become richer." In summary, Oliver-Smith contends that Haitian elites "impoverish[ed] the population with brutality, militarism, mismanagement, and corruption," while doing "little to construct a viable infrastructure of institutional framework for the country."

As Clair Antoine Payton (2021) has pointed out, Baby Doc's culture of corruption contributed directly to the death toll in the 2010 quake through the medium of concrete. In the early years of Baby Doc's rule, Haiti was gripped with a new optimism, fueling a building boom. Many of this era's structures were built out of concrete, in large part because wood was scarce in heavily deforested Haiti. What is more, concrete had its advantages: not only was it resistant to hurricanes, it had cultural significance for Haitians, as it symbolized modernity, stability, social mobility, and "protection from what they perceived as a violent and unpredictable society." However, due to crony capitalism in Haiti, the Duvalier regime maintained a monopoly over cement, artificially inflating its price. As a result, when Haitian builders mixed their concrete, they skimped on the costly cement, producing concrete that was full of sand and gravel but had as little as one quarter of the minimally recommended amount of cement binding material.

Making matters worse, this was a period of uncontrolled growth in the capital city, Port-au-Prince. In the absence of enforceable zoning laws and building codes, recent Haitian migrants from the countryside constructed flimsy homes on whatever land was available, including swampy shorelines and steep slopes. Reporter Amy Wilentz described such housing as "seemingly made of plaster and glue and paper clips and washboards and oil drums and pieces of cardboard and tin" (Stengel et al., 2010). Haiti's "monstrous urbanization", to borrow a phrase from Mark Schuller (2016), continued even after the fall of Baby Doc. The population of Haiti's capital quadrupled between 1980 and 2010, triggering the Haitian equivalent of "white flight" and the slummification of much of Port-au-Prince. Many of these new shanty town neighborhoods lacked electricity, running water, and even serviceable roads, which would become a problem during the rescue and recovery stage of the quake.

Haiti's vulnerability to the 2010 earthquake was heightened by other factors as well. Haiti's economy was in free fall in the years before the quake, due in part to neoliberal reforms that were supposed to make Haiti a center of industry but which instead decimated Haiti's agricultural sector. Unable to compete against subsidized food imports from abroad, many Haitian farmers abandoned their land and left for the cities, further swelling the already-swollen population of Port-au-Prince. Much of Haiti's declining revenues continued to be eaten up by foreign debts. Disease contributed to Haiti's poverty as well; not only was Haiti the Western Hemisphere epicenter of the AIDS epidemic, Haiti caved in to foreign pressure and slaughtered most of its pigs during the swine flu scare of 2009–2010, further impoverishing an already destitute population. Even before the arrival of the 2010 quake, 78% of Haitians lived on less than $2/day and fully 81% of the population was malnourished. Making matters still worse, chronic instability after the fall of the Duvaliers in Haiti provoked the United Nations into sending a permanent peacekeeping force, MINUSTAH, in 2004. These peacekeepers would eventually be victims of the quake themselves, but also the causal agents of a horrific post-quake cholera epidemic.

The 2010 Quake and Aftermath: Goudougoudou and Kolera

The 2010 quake lasted only 35 seconds, but for those who felt its seismic waves, it seemed to last a lifetime. According to reporter Nancy Gibb, "it was the kind of quake that rings church bells in steeples before they collapse, makes buildings heave and sway like ships in the sea, turns stone to powder, wrinkles walls like tissue paper" (Stengel et al., 2010) The Haitians would call the quake the *doz*, the *bagay la* (the thing), *evenman nan* (the event), and/or the *goudougoudou*, an onomatopoeic term referring to the sound made by the earthquake.

By whatever name, the quake inflicted horrific destruction upon Port-au-Princes' vulnerable urban landscape. The quake destroyed an estimated 25,000 public buildings, including 21 of 22 of Haiti's government ministries. Haitian officials would later claim that 250,000 private residences and 30,000 commercial buildings had collapsed or were so damaged as to require post-quake demolition. As might be imagined, the death toll was atrocious. When Haitian author Yanick Lahens ventured into Port-au-Prince the day after she quake, she crossed

> a chaotic city, littered with corpses, some already covered with a white cloth or a simple cardboard box, children's bodies, young people stacked in front of schools, flies already dancing around other corpses, injured people, haggard older people, destroyed building and houses. All that was missing were the trumpets of the Angel of Apocalypse to announce the end of the world.
>
> (Munroe, 2010)

In the aftermath of the quake, the Haitian government estimated that the eventual death toll might have reach 316,000, surpassing that of the Indian Ocean tsunami six years earlier. However, this high number has not been supported by subsequent studies, who suspect that the Haitian government may have exaggerated the deaths in an attempt to appeal to international aid donors. A 2010 University of Michigan study estimated the casualties at 160,000, while the UN put the number at between 46,000 and 85,000. The reality is that we don't know how many died, in part because of weak record-keeping by the dysfunctional Haitian state. As for the living, fully one-sixth of Haiti's whole population was forced into refugee camps after the quake, where they suffered from poor sanitation, lack of running water, and in many cases sexual violence and exploitation.

Suffering on this scale provoked a generous response from the international humanitarian community, though as is typical of the work of "humanitarian international," only a limited amount of the donations reached the people in need. According to a study by Vijaya Ramachandran and Julie Walz (2015), only 1% of the donated aid paid for immediate humanitarian assistance, while 15–20% was soaked up by a Haitian government with a well-earned reputation for corruption. The bulk of the rest went to NGO's (non-governmental organizations) and private contractors, mostly from outside Haiti, who built an "extensive infrastructure for the provision of social services" but coordinated with neither the government or each other and spent their money without any real accountability. As a result, Ramachandran and Walz argue, "almost $700 was spent per Haitian citizen in the two years since the 2010 quake," but there is little to show for it, prompting some critics to speculate that Haiti would have been better off if relief organizations had just given $700 per person directly to the Haitians.

For Justin Podur (2012), the story that symbolizes the aid community's post-quake mismanagement involves the estimated 19 million cubic meters of rubble produced by the earthquake. The logical thing would have been to employ out-of-work Haitian survivors to remove the rubble, but instead contracts for rubble removal were awarded to US, Canadian, and elite Haitian firms, which brought in their labor from outside of the country. As a result, "the bulk of the Haitian urban population sat there, unemployed or engaged in petty economic activities and watched … [as] dump trucks gathered up and brought in the rubble." For unemployed Haitians, "that rubble could have been gold," and an opportunity to offer a "massive economic stimulus" for poor Haitians was missed.

Given the scale of the tragedy in Haiti, and the ubiquity of disease cues like blood, death, and decay, it is somewhat surprising that there were few outbreaks of xenophobic violence in the aftermath of the quake. This is not to say that violence did not occur; both before and after the quake, gang violence and turf wars were chronic in Haiti's shantytowns. While there was an uptick of violence after the quake, it did not reflect xenophobia so much as the opportunistic escape of 4,000 prisoners from Port-au-Prince's jails, including many gang leaders who returned to their neighborhoods and tried to re-take power from former "lieutenants" who had seized control in their absence. What is more, there was a post-quake spike in sexual violence in the crowded refugee camps due to the breakdown of normal extended family and neighborhood social structures.

About nine months after the quake, however, some Haitians were driven to xenophobic violence by another catastrophe, a massive cholera epidemic that began on the banks of Haiti's Artibonite River. *Kolera*, as the disease was called in Haitian creole, eventually infected over 800,000 Haitians and resulted in over 9,600 deaths. While some international observers believed the cholera epidemic might have been triggered by the quake itself, perhaps by altering the marine ecosystem of the Caribbean Sea, we now know that it likely originated in a MINUSTAH camp of Nepali peacekeepers. However, the UN tried to cover up this fact, fearing that foreign peacekeepers might be scapegoated for the disease. While some Haitians made the link between *kolera* and the peacekeeping force, other Haitians found a different scapegoat for the epidemic: Haitian Vodou priests who were accused of poisoning wells with sinister magic powders. In the end at least 45 of these Vodou priests were lynched by Haitian mobs, and it is likely they were singled out for violence in part because Vodou was tainted by its association with the hated Duvalier regime.

As I write this in 2023, Haiti still has not recovered meaningfully from the earthquake of 2010. Indeed, in a country like Haiti, it is hard to imagine what "recovery" would look like, given that domestic political oppression and corruption, rampant gang violence, chronic debt, and economic disadvantage were the norm even before the quake. As Oliver-Smith (2012) reminds us, disasters are processes, not isolated events, and "much of the devastation and misery caused in Haiti by the earthquake of January 12" was the generated by "historical processes set in motion since the time of independence, or even earlier." In the final analysis, Oliver-Smith contends, the quake was not just a natural disaster, but a "socially created phenomenon" – a basic truth that applies not only to Haiti's 2010 quake, but also to the 1755 Lisbon Earthquake, the 1906 San Francisco quake, and the 1923 Kantō Plain Earthquake as well.

Further Reading

Albini, Paola and Kenji Satake. "Voices of Foreign Residents in Yokohama and Tokyo at the Time of the 1923 Kanto Earthquake." *Journal of Disaster Research*, vol. 18, no. 6 (2023), pp. 598–610.

Bilham, Roger. "Lessons from the Haiti Earthquake." *Nature*, vol. 463, no. 18 (2010), pp. 878–879.

Chester, David K. "The 1755 Lisbon Earthquake." *Progress in Physical Geography*, vol. 25, no. 3 (2001), pp. 363–383.

Frerichs, Ralph R. *Deadly River: Cholera and Cover-Up in Post-Earthquake Haiti.* London: ILR Press, 2016.

Glanz, James and Norimitsu Onishi. "Japan's Strict Building Codes Saved Lives." *New York Times*, March 11, 2011.

Gupta, Harsh K. and Vineet K. Gahalaut. *Three Great Tsunamis: Lisbon (1755), Sumatra-Andaman (2004), and Japan (2011)*. New York: Springer, 2013.

Hammerl, Christa. "Earthquakes." In Shephard Krech III, J. R. McNeill, and Carolyn Merchant (Eds.), *Encyclopedia of World Environmental History*, vol. 1. London: Routledge, 2003, pp. 358–364.

Hobbs, W. H. "The Messina Earthquake." *Bulletin of the American Geographical Society*, vol. 41, no. 7 (1909), pp. 409–422.

Howe, Maud. *Sicily in Shadow and Sun: The Earthquake and the American Relief Work*. Boston, MA: Little, Brown, and Company, 1910.

Jones, Lucy. *The Big Ones: How Natural Disasters Have Shaped Us (and What We Can Do About Them)*. New York: Anchor Books, 2018.

Kenji, Hasegawa. "The Massacre of Koreans in Yokohama in the Aftermath of the Great Kanto Earthquake of 1923." *Monumenta Nipponica*, vol. 75, no. 1 (2020), pp. 91–122.

Kutchment, Anna. "Even if Injection of Fracking Wastewater Stops, Quakes Won't." *Scientific American*, September 9, 2019.

Marcelin, Louis Herns. "Violence, Human Insecurity, and the Challenge of Rebuilding Haiti: A Study of a Shantytown in Port-au-Prince." *Current Anthropology*, vol. 56, no. 2 (2015), pp. 230–255.

Martins, A. Nuno, Catherine Forbes, Andreia Amorim Pereira, and Daniela Matos. "The Changing City: Risk and Built Heritage. The Case of Lisbon Downtown." *Procedia Engineering*, vol. 212 (2018), pp. 921–928.
Oishi, Shigehiro and Asuka Komiya. "Natural Disaster Risk and Collectivism." *Journal of Cross-Cultural Psychology*, vol. 48, no. 8 (2017), pp. 1263–1270.
Poirier, Jean-Paul. "Electrical Earthquakes: A Short-Lived Theory in the 18th Century." *Earth Sciences History*, vol. 35, no. 2 (2016), pp. 283–302.
Poole, Otis M. *The Death of Old Yokohama: In the Great Japanese Earthquake of 1923*. London: Routledge, 2010.
Prager, Ellen. *Furious Earth: The Science and Nature of Earthquakes, Volcanoes, and Tsunamis*. New York: McGraw-Hill, 2000.
Reilly, Benjamin. *Disaster and Human History: Case Studies in Nature, Society, and Catastrophe*. 2nd ed. Boone, NC: McFarland, 2022.
Schencking, Charles J. "The Great Kanto Earthquake and the Culture of Catastrophe and Recovery in 1920's Japan." *Journal of Japanese Studies*, vol. 34, no. 2 (2008), pp. 295–331.
Sekizawa, Ai. "1923 Great Kanto Earthquake: Fire Damage and Lessons Learned." *Journal of Disaster Research*, vol. 18, no. 6 (2023), pp. 558–561.
Sng, Oliver, Stephen L. Neuberg, Michael E. W. Varnum, and Douglas T. Kenrick. "The Crowded Life Is a Slow Life: Population Density and Life History Strategy." *Journal of Personality and Social Psychology*, vol. 112, no. 5 (2017), pp. 736–754.
Sykes, Lynn R. *Plate Tectonics and Great Earthquakes: 50 Years of Earth-Shaking Events*. New York: Columbia University Press, 2019.
Tobriner, Stephen. "La Casa Baraccata: Earthquake-Resistant Construction in 18th-Century Calabria." *Journal of the Society of Architectural Historians*, vol. 42, no. 2 (1983), pp. 131–138.
Wheatcroft, Andrew. *Infidels: A History of the Conflict Between Christendom and Islam*. New York: Random House, 2005.

Works Cited

Allen, Michael J. "The Price of Identity: The 1923 Kantō Earthquake and its Aftermath." *Korean Studies*, vol. 20 (1996), pp. 64–93.
Bankoff, Greg. "Design by Disasters: Seismic Architecture and Cultural Adaptation to Earthquakes." In Fred Krüger*et al.* (Eds.), *Cultures and Disasters: Understanding Culture Framings in Disaster Risk*. London: Routledge, 2015, pp. 53–71.
Bellegarde-Smith, Patrick. "A Man-Made Disaster: The Earthquake of January 12, 2010 – A Haitian Perspective." *Journal of Black Studies*, vol. 42, no. 2 (2011), pp. 264–275.
Bentzen, Jeanet Sinding. "Acts of God? Religiosity and Natural Disasters Across Subnational World Districts." *The Economic Journal*, vol. 129 (2019), pp. 2295–2321.
Davies, Andrea Rees. *Saving San Francisco: Relief and Recovery after the 1906 Disaster*. Philadelphia, PA: Temple University Press, 2012.
Donlon, Rachel A. (Ed.) *Haiti Earthquake and Response*. New York: Nova Science Publishers, 2012.
Dyl, Joanna L. *Seismic City: An Environmental History of San Francisco's 1906 Earthquake*. Seattle, WA: University of Washington Press, 2017.

Fradkin, Philip L. *The Great Earthquake and Firestorms of 1906*. Berkeley, CA: University of California Press, 2005.

Hagan, Ryan. "Acts of God, Man, and System: Knowledge, Technology, and the Construction of Danger." In Jacob A. C. Remes and Andy Horowitz (Eds.), *Critical Disaster Studies*. Philadelphia, PA: University of Pennsylvania Press, 2021, pp. 32–50.

Hammer, Joshua. *Yokohama Burning: The Deadly 1923 Earthquake and Fire That Helped Forge the Path to World War II*. New York: Free Press, 2006.

Hanska, Jussi. *Strategies of Sanity and Survival: Religious Responses to Natural Disasters in the Middle Ages*. Helsinki: Finnish Literature Society, 2002.

Kendrick, T. D. *The Lisbon Earthquake*. New York: J. B. Lippincott, 1955.

Kierner, Cynthia A. *Inventing Disaster: The Culture of Calamity from the Jamestown Colony to the Johnstown Flood*. Chapel Hill, NC: University of North Carolina Press, 2019.

Klein, Naomi. *The Shock Doctrine: The Rise of Disaster Capitalism*. London: Picador, 2008.

Lee, Nihnee. "The Enemy Within: Earthquake, Rumors, and Massacre in the Japanese Empire." *Faculty Research and Creative Activity*, vol. 58 (2008), pp. 187–211.

Molesky, Mark. "The Vicar and the Earthquake: Conflict, Controversy, and a Christening during the Great Lisbon Disaster of 1755." *E-Journal of Portuguese History*, vol. 10, no. 2 (2012), pp. 76–94.

Molesky, Mark. *This Gulf of Fire: The Destruction of Lisbon*. New York: Alfred A. Knopf, 2015.

Munroe, Martin (Ed.) *Haiti Rising: Haitian History, Culture and the Earthquake of 2010*. Kingstown, Jamaica: University of the West Indies Press, 2010.

Oliver-Smith, Anthony. "Haiti's 500-Year Earthquake." In Mark Schuller and Pablo Morales (Eds.), *Tectonic Shifts: Haiti Since the Earthquake*. Boulder, CO: Kumarian Press, 2012.

Oliver-Smith, Anthony. "Conversations in Catastrophe: Neoliberalism and the Cultural Construction of Disaster Risk." In Fred Krüger *et al.* (Eds.), *Disasters and Cultures*. New York: Routledge, 2015, pp. 37–52.

Paice, Edward. *Wrath of God: The Great Lisbon Earthquake of 1755*. London: Quercus, 2008.

Palmer, James. *Heaven Cracks, Earth Shakes: The Tangshan Earthquake and the Death of Mao's China*. New York: Basic Books, 2012.

Payton, Clair Antone. "Concrete Kleptocracy and Haiti's Culture of Building: Towards a New Temporality of Disaster." In Jacob A. C. Remes and Andy Horowitz (Eds.), *Critical Disaster Studies*. Philadelphia, PA: University of Pennsylvania Press, 2021, pp. 71–84.

Pfister, Christian. "'The Monster Swallows You': Disaster Memory and Risk Culture in Western Europe, 1500–2000." *Rachel Carson Center Perspectives*, no. 1 (2011), pp. 1–23.

Podur, Justin. *Haiti's New Dictatorship: The Coup, the Earthquake, and the UN Occupation*. New York: Pluto Press, 2012.

Ramachandran, Vijaya and Julie Walz. "Haiti: Where Has All the Money Gone?" *Journal of Haitian Studies*, vol. 21, no. 1 (2015), pp. 26–65.

Ryang, Sonia. "The Great Kanto Earthquake and the Massacre of Koreans in 1923: Notes on Japan's Modern National Sovereignty." *Anthropological Quarterly*, vol. 76, no. 4 (2003), pp. 731–748.

Schuller, Mark. *Humanitarian Aftershocks in Haiti*. New York: Rutgers University Press, 2016.

Seidensticker, Edward. *Low City, High City: Tokyo from Edo to the Earthquake*. New York: Alfred A. Knopf, 1983.

Shrady, Nicholas. *The Last Day: Wrath, Ruin and Reason in the Great Lisbon Earthquake of 1755*. New York: Penguin Books, 2008.

Sloan, Doris. *Geology of the San Francisco Bay Region*. Oakland, CA: University of California Press, 2006.

Smits, Gregory. "Shaking up Japan: Edo Society and the 1855 Catfish Picture Prints." *Journal of Social History*, vol. 39, no. 4 (2006), pp. 1045–1078.

Smits, Gregory. *Seismic Japan: The Long History and Continuing Legacy of the Ansei Edo Earthquake*. Honolulu, HI: University of Hawaii Press, 2012.

Stengel, Richard, Michael Elliot, Jefferey Kluger, and Richard Lacayo (Eds.) *Earthquake Haiti: Tragedy and Hope*. New York: Time Books, 2010.

van Bavel, Bas *et al*. *Disasters and History: The Vulnerability and Resilience of Past Societies*. New York: Cambridge University Press, 2020.

Vermij, Rienk. "Subterranean Fire: Changing Theories of the Earth During the Renaissance." *Early Science and Medicine*, vol. 3, no. 4 (1998), pp. 323–347.

Vermij, Rienk. *Thinking on Earthquakes in Early Modern Europe*. London: Routledge, 2020.

Voltaire. *Candide: or, Optimism*. New York: Penguin Books, 1947.

Wisner, Ben, Piers Blaikie, Terry Cannon, and Ian Davis. *At Risk: Natural Hazards, People's Vulnerability and Disasters*. 2nd ed. New York: Routledge, 2004.

Wisner, Ben, J. C. Gaillard, and Ian Kelman (Eds.) *The Routledge Handbook of Hazards and Disaster Risk Reduction*. New York: Routledge, 2011.

3 Tsunamis

On December 26, 2004, Thai fish-seller Ratree Kaewaied was waiting for a job interview in the lobby of the under-construction Bang Sak Beach Hotel when she realized that something was wrong. In all her life by the seashore, she had never seen the tide so low. The ground seemed to be shaking, she remembered, and she heard a "rattling, chugging noise" like a train, though she was far from any tracks. Seconds later, Ratree was running for her life from an enormous oncoming wave, which in her panicked mind took on the guise of a "devil hand reaching out to grab her."

And grab her it did. Ratree was catapulted face-forward into the sand and then engulfed by the surging water, which tumbled and pummeled her for two minutes before letting her return to the surface. No sooner had she reached her feet than she was sucked under again, and again, draining her strength. She finally found herself pinned against the side of a lobby staircase by the force of the water. Ratree, whose battered body by this point was "radiat[ing] with pain," would probably have drowned there if not for the intervention of an anonymous rescuer who took her arm and pulled her atop the lobby stairs.

No sooner had she reached the relative safety of the stairs than a second wave, larger than the first, began to surge into the hotel lobby. Ratree and her rescuer scrambled onto a tower of scaffolding in advance of the wave. From that perch, they watched in horror as a jumble of building debris, cars, boats, and struggling people was carried far inland before being sucked back out to sea. Then a third wave washed ashore, and a fourth. It was as if, she thought, the ocean water had become the tongue of a massive frog, "lashing out to capture its prey and then sucking it back in."

After the fourth wave receded, Ratree climbed down to the ground and picked her way over the muddy ground, which was strewn with hundreds of twisted bodies. Desperate to return to her family in the village of Nam Khem, she headed north along what she thought was the highway, nearly invisible beneath a layer of debris. Everywhere she looked there were dead bodies, "faces protruding from the mud and

DOI: 10.4324/9781003436805-4

peering out of the windshields of cars that had been twisted into knots." As she worked her way north, she hoped against hope that her home in Nam Khem had been untouched and that the waves had limited their fury to Bang Sak Beach.

Ratree would later find out that the tsunami was bigger than Bang Sak and even bigger than Thailand. On the day after Christmas of 2004, the entire Indian Ocean basin reeled from a disaster on a scale that, as I write this in 2023, has not yet been surpassed in the twenty-first century. What is more, in relative terms, Thailand actually got off pretty easy. Thailand suffered only a small fraction of the much higher death tolls reported in Sri Lanka and Indonesia.

This fact would probably have been little comfort to Ratree, however. When she finally got home, she would learn that the young daughter that she had left behind that morning in Nam Khem had been lost to the Indian Ocean Tsunami.

The Science of Tsunamis

First, a note on naming conventions. Since Britain has long experienced extreme tidal variation but only weak and infrequent tectonically-induced waves, the English language has long lacked a good term for the latter. In the absence of something better, such events have generally been called "tidal waves" in English, which was misleading as tsunamis have nothing to do with the tides. In contrast, the Japanese have had a specialized word for these waves since at least 869 CE. In that year, Japan was struck by the Jōgan Earthquake, a tremendous tremor on the same scale as the 2011 Tōhoku Earthquake described later in this chapter. Fishermen who had departed from intact villages on the morning of this earthquake returned in the evening to a coastline rendered almost unrecognizable, ravaged by a flood so great that "you could not tell where the sea ended and the land began." Horrified by the death and destruction in coastal harbors, but puzzled by the fact that they had noticed no particularly large waves out at sea, the fishermen and their families labeled the event a tsunami, or "harbor wave." This Japanese term first passed into English following a *National Geographic* article in 1896, though it did not become commonplace until after the 2004 Indian Ocean Tsunami described in the case studies below.

Tsunamis are usually (about 90% of the time) the result of plate displacement due to undersea earthquakes (Figure 3.1). The other 10% of tsunamis are the product of various other processes. As we will see in Chapter 4, volcanoes can cause tsunamis by displacing the ocean's surface with ejected debris or pyroclastic flows, triggering tsunami-style waves. Similarly, the impact of extraterrestrial objects can trigger powerful tsunamis. Around 35 million years ago, for example, the east coast of the United States was struck by a huge comet or asteroid, leaving a crater 38 kilometers across that helped to form Chesapeake

Figure 3.1 Victims of the tsunami that accompanied the 1896 Maiji Sanriku earthquake
Source: University of British Columbia Library, Rare Books and Special Collections.

Bay. The resulting tsunami may have reached as far inland as the Blue Ridge Mountains, over 350 kilometers away. In addition, some tsunamis are the result of terrestrial or submarine landslides, which are typically though not necessarily triggered by earthquakes. Since the strength of a tsunami is relative to the displacement of water, and since a landslide can shift a huge amount of material in a short span of time, landslides can trigger truly stupendous tsunamis. Case in point was the 1958 Lituya Bay Earthquake in Alaska, during which rockfall from a mountain face striking the waters of the bay generated a "mega-tsunami" that may have reached over 520 meters in height. Luckily, Lituya Bay had few inhabitants, so the tsunami led to only moderate damage and five reported casualties.

Whatever their cause, tsunami waves have little in common with the ocean waves that we see on a day at the beach. Ocean waves are generally the result of the friction between wind and the surface of the water, which stacks the water atop itself in crests. Such waves can travel long distances, but they are not nearly as powerful as tsunamis, since the energy transfer is confined to the surface layer of the ocean. In contrast, the energy of a tsunami derives from the displacement of a column of water, for example, when a section of the seafloor rises up or drops down during an earthquake. Since the energy of a tsunami wave is contained by the entire column of water above the displacement zone, tsunamis are potentially far more powerful than their wind-driven ocean kin.

Although the initial energy leading to water displacement generally derives from tectonic forces, the actual spread of tsunami waves from their point of origin is caused by gravity, which pushes the upraised column of water down and outwards or, if the original direction of displacement was downward, pushes water from both sides of the displacement zone into the void created by the descending water column. To understand this better, imagine plopping down into a tub full of water. While you caused the original displacement of the water with your body, the subsequent back-and-forth sloshing of the water in the tub is due to gravity, which is pushing the disturbed surface back into a state of equilibrium.

Tsunami waves are therefore gravity waves, and like all wave-like phenomena, they have both a wavelength (the distance between crests) and an amplitude (the vertical displacement from the undisturbed neutral position, which in the case of tsunami waves is sea level). Unlike regular waves, tsunami waves have a low amplitude but an extremely long wavelength, sometimes spanning hundreds of kilometers, which is why the fishermen at the time of the 869 Jōgan tsunami took no notice as the tsunami waves passed under their boats. When this gravity wave moved into the shallows near Japan's harbors, some of this energy was lost to friction with the seafloor, but the remaining energy was confined to a progressively smaller column of water. The tsunami's wavelength therefore decreased, but as it did, the amplitude of the wave had to increase accordingly. As a result, the same wave that passed harmlessly under the fishermen swelled to a height of 6 or 7 meters along the coastline, reportedly killing 1,000 coastal residents in the process.

While tsunami waves, like all waves, lose energy as they travel from their point of origin, the relatively long wavelength of tsunami waves means that they can transmit energy over great distances. The waves from the 2011 Tōhoku Tsunami off the coast of Japan crossed 17,000 kilometers of the Pacific Ocean to Chile, where crests reached 2 meters in height and inflicted serious damage on the Chilean scallop industry. That being said, damage and loss of life tend to be greatest close to the zone of displacement, which is generally the epicenter of an earthquake.

In the case of such earthquake-generated tsunamis, the potential strength of the tsunami is closely tied to the type of fault boundary. Tsunamis are generally not created by transverse faults, where the movement is lateral along the plate boundaries, as movement along these faults typically does not lead to vertical displacement (though it could trigger landslides). Frequent tsunamis are produced by normal faults, such as along mid-oceanic ridges, though the constant release of energy along these fault lines prevents the build-up of tension that could lead to strong quakes. As a result, the fault boundaries that are most associated with tsunamis are thrust faults, particularly subduction faults, where an oceanic plate is diving beneath a continental plate. When plate movement occurs on such a fault, it typically takes the form of horizontal plate displacement, and since the two plates may have been locked together for centuries, the built-up tension can be considerable, sometimes generating a quake in excess of 9.0 moment magnitude quake. The fact that subduction zone earthquakes typically occur offshore, beneath a column of seawater, compounds the problem for nearby inhabitants.

These tectonic factors ensure that the degree to which a given coastline is susceptible to large tsunamis is roughly proportional to that region's proximity to an undersea subduction zone. The largest such zone is along the rim of the Pacific oceanic plate, especially in the eastern Pacific, where the Pacific plate is diving under the Eurasian, North American, Philippine, and Okhotsk Plates. Plate movement along this zone generates powerful tsunamis, and also strong quakes, as we saw in the section on the Great Kantō Earthquake in Chapter 2. A second major subduction zone lies beneath the Mediterranean Sea, where the Africa plate is diving under the Eurasian plate. According to Gerassimos Papadopoulos, the tsunamis generated by this plate movement have frequently punctuated Greek history, and may have even seeped into Homer's *Iliad*, which mentions a severe and unexpected ocean wave that swamped the camp of the Greeks besieging Troy. Another Mediterranean tsunami struck the Egyptian city of Alexandria in 365 CE – about 50,000 homes and 5,000 lives were lost to a wave that may have reached 30 meters in height, an event so traumatic that Alexandria remembered this "day of horror" with an annual commemoration as late as the sixth century. Finally, as we will see in the Indian Ocean Tsunami case study below, the subduction of the India-Australia Plate beneath the Eurasia Plate in the eastern Indian Ocean can also generate powerful tsunamis. However, such tsunamis are comparatively infrequent due to the slow relative speed of the colliding plates.

Tsunamis and Culture

As might be expected, regions of the coast that are regularly struck by tsunamis have developed "cultures of disaster" in response to the threat, though to different degrees. One limiting factor in the development of such a culture, when it comes to tsunamis, is the sheer scale of these events: other than abandonment of the coastline, either temporarily or permanently, there was little that a traditional society with limited technological means could do to mitigate tsunami damage. The development of a culture of disaster was also limited, in some areas, by the infrequency of tsunamis, which means they are more easily forgotten and less likely to be incorporated into cultural memory and practices.

Cultures of disaster surrounding tsunamis can be divided into two broad categories: (1) beliefs about the origins and meanings of these events, and (2) more practical measures taken by societies to mitigate death and destruction. Turning first to the issue of origin and meanings, Europe's culture of tsunamis was notable by its absence, which reflects Europe's relative unfamiliarity with such events. When a major tsunami did occur, Europeans would interpret it as an unprecedented and extraordinary event which demanded an extraordinary explanation. The tsunami that accompanied the Lisbon Earthquake of 1755, for example, was seen as the fulfillment of scripture, specifically Revelations 16:3–4, which declared that at the end of days an angel would "pour out his vial upon the sea, and it became as the blood of a dead man: and every living soul died in the sea."

Japanese culture also linked tsunamis to divine intervention, though in the case of Japan tsunamis were seen as expected occurrences in the normal order of things rather than signs of the world's imminent end. The Japanese associated tsunamis with earthquakes, and earthquakes to celestial beings, including the *ryūjin*, dragons that lived deep beneath the sea. These *ryūjin*, incidentally, were probably the Japanese version of the *nāga* serpents of Indian mythology, who came to Japan via Buddhist texts. The medieval Japanese believed that the movement of these monsters were acts of punishment, triggered by divine anger about human violation of taboos or religious prohibitions. By the sixteenth century, however, these earlier ideas linking divine anger to disasters had been subsumed into the Chinese "mandate of heaven" theory, which linked disaster to bad and unethical governmental leadership and which regarded natural disasters like earthquakes, famines, and floods as the "automatic karmic consequences of the Buddhist deities withdrawing their protection against negative forces." Such beliefs were still in force during the nineteenth and twentieth centuries, though the *ryūjin* sea dragons had morphed over time into whales or fish, including the quake-causing *Namazu* catfish we met in Chapter 2. Indeed, the idea that natural disasters represent divine punishment still exists today; as Fabio Rambelli (2014) points out, after the 2011 Tōhoku Tsunami the governor of Tokyo ascribed that disaster to heavenly displeasure over Japan's over-attachment to the US and its "facile and passive politics of peace."

In terms of practical tsunami mitigation measures, different societies in tsunami-prone regions have incorporated tsunami preparations into their cultural practices, though to varying degrees. This disparity in tsunami preparedness comes out clearly in the comparative study by Esteban et al. (2013) into the "tsunami cultures" of Chile, Indonesia, and Japan. Despite their proximity to a subduction zone, Esteban et al. found that the Chilean population has done little to prepare for a tsunami. Residents know to run for higher ground when threatened by a tsunamis, which were locally known as *maremotos* or sea quakes, but Esteban et al. found entire villages built "without any consideration … of countermeasures" within the vulnerable coastal zones. Esteban et al. do not speculate why Chileans were so unprepared, but one possibility is that Chilean culture is to a large degree a transplanted European culture, and, as discussed above, tsunamis were overall a rare occurrence in Europe. In contrast, the Indonesian populations they surveyed were somewhat better prepared for tsunamis. Like the Chileans, Indonesians knew to run inland after a quake; unlike the Chileans, their villages had been built somewhat inland from the sea and many were shielded by a protective layer of forest. However, it was unclear to these researchers whether this was the result of active planning or just the passive consequence of the progressive abandonment of coastal sites which had fallen victim to earlier tsunamis. Given the relative infrequency of Indian Ocean tsunamis, this latter explanation may be the more likely one.

Not surprisingly, Esteban et al. found that the Japanese had by far the best anti-tsunami preparations, which they ascribe to both higher levels of resources in Japan as well as the relative frequency of these events. Japan's culture of disaster surrounding tsunamis was an ancient one, inscribed into the landscape itself in the form of tsunami memorials, standing stones and trees erected at the high-water mark of previous tsunami disasters. Knowing that their past was also their future, Japanese living on the coast of Tōhoku had taken numerous steps to mitigate the tsunami hazard, including early detection systems, evacuation drills, abandonment of the vulnerable coastal plains, and even the construction of expensive seawalls to protect harbor towns. As we will see in the case study below, however, these preparations were not sufficient to fully protect the Tōhoku region from the ravages of the powerful 2011 earthquake and tsunami.

The Sumatra-Andaman Earthquake and the Indian Ocean Tsunami, 2004

Seven years before the Tōhoku catastrophe, an even more powerful earthquake struck a region that was far less prepared than Japan for a tsunami, triggering the worst natural disaster so far in the twenty-first century. The disaster began with a rip and a tear on December 26, 2004 in a tectonically complicated region just offshore of Sumatra. Here, the Indo-Australian Plate is diving beneath the Eurasian Plate, or more technically the Burma Plate, a sub-region of the Eurasian tectonic plate. What is more, the Indo-European Plate is itself fracturing at this point, slowly dividing itself into an Indian Plate and an Australian Plate, which are drifting in the same direction but at very different velocities. Since the Indo-Australian Plate is moving at only half the speed or less as the Pacific Plate, earthquakes and tsunamis are less common in the Indian Ocean than the eastern Pacific. However, since this plate juncture is a subduction zone, these quakes, though infrequent, can be extremely powerful. The 2004 Sumatra-Andaman earthquake that caused the Indian Ocean Tsunami, in fact, was at least the third most powerful earthquake ever recorded by modern seismographic instruments, with a moment magnitude somewhere between 9.0 and 9.3.

Making matters worse is the sheer length of the fault line that slipped during the quake. While the original epicenter of the quake lay about 100 kilometers from Sumatra and 50 km from the smaller island of Simeuluë, movement along the plate continued north for 1,600 kilometers over the course of 10 minutes, nearly reaching the coast of Myanmar. Along this entire line, the seabed suddenly lifted by several meters, suddenly displacing an estimated 30 cubic kilometers of ocean water. Because the fault line was so long, it diffused tsunami waves like a long florescent bulb diffuses light, ensuring that the resulting tsunami would penetrate almost every corner of the Indian Ocean basin by reducing the protective effect of headlands and peninsulas. As a result, the Indian Ocean Tsunami was a truly international calamity, claiming lives and property in 14 different nations.

As is typical of tsunamis, the height of the wave crests depended in large part on the distance from the displacement zone, though with two important caveats. First, movement along the northern end of the fault line was weaker than to the south near the epicenter of the quake; had this not been the case, Bangladesh would have been struck by a 5–6 meter tsunami that could conceivably have killed a million people in this low-lying nation. Second, the fact that the rupture zone ran south to north, and thus propagated tsunami waves mainly to the east and west, meant that Myanmar received much lower tsunami crests than would otherwise have been expected, given that Myanmar lies just a few hundred kilometers north from the endpoint of the rupture zone. Myanmar thus got off relatively lightly, though exactly how lightly is a matter of conjecture, since the tight-lipped military regime that ruled Myanmar in 2004 was not very forthcoming about the scale of the disaster it suffered. By the same token, regions that were directly east and west of the rupture zone received the brunt of the tsunami's impact and recorded by far the most casualties, including portions of Indonesia, Sri Lanka, India, and Thailand.

In all, the Indian Ocean Tsunami may have claimed as many as 228,000 victims, but this bland overall death toll is not very helpful in understanding how the tragedy unfolded in particular locations. While the Indian Ocean Tsunami represented a general hazard to the entire Indian Ocean basin, Karan and Shanmugan (2011) have argued compellingly that this hazard only became a disaster when it interacted with vulnerable populations. Vulnerable demographic groups such as "women, children, the elderly, and the disabled," were worse impacted by the tsunami. So too were economically disadvantaged groups, who often occupied inferior housing in flood-prone terrain and who lacked the resources to recover quickly from the disaster. Other vulnerabilities were inherent in specific local socioeconomic, political, geographical, cultural, and other circumstances. Since these factors are complex, the rest of this section will try to do them justice by focusing on the impact of the Indian Ocean Tsunami in only two nations, Indonesia and Sri Lanka, which between them accounted for nearly 90% of the overall death toll.

The Historical Roots of Vulnerability in Aceh

The Indonesian territory most impacted by the Indian Ocean Tsunami was Aceh (pronounced ah-che), Indonesia's westernmost province, which occupies the northern tip of the island of Sumatra. Aceh has several claims to fame. For one thing, Aceh was the first part of Indonesia to adopt Islam, which was introduced by Arab Muslim merchants in the mid-thirteenth century, and even today Aceh has a reputation for Islamic piety. Aceh was also the site of a powerful sixteenth–seventeenth-century sultanate, which competed with the Portuguese and regional rivals for control over the lucrative trade passing through the nearby Malacca Strait. Aceh's power waned in the eighteenth century, in part due to defeats at the hands of Portugal and its allies, but waxed again in the nineteenth century, when a reconstituted Aceh Sultanate controlled the northern half of Sumatra, held the vassalage of much of the southern Malay Peninsula, and controlled the production of fully half of the world's black pepper.

Unfortunately, Aceh's wealth attracted the attention of the Dutch, who by the mid-nineteenth century had already consolidated their control over Java, Borneo, and much of southern Sumatra. Troops from the Dutch East Indies invaded Aceh in 1873 and soon occupied the coastal cities, but stubborn Acehnese guerilla warfare ensured that the Dutch would not exert complete control over Aceh until well into the twentieth century. Aceh was then occupied by Japan during World War II, once again provoking fierce Acehnese resistance which was bloodily suppressed. Nor did Aceh accept the control of the independent state of Indonesia, which was formed shortly after Japan's defeat in 1945. Despite promises of autonomy, including the right to practice Islamic *shari'a* law, Aceh was incorporated into the Indonesian province of North Sumatra in 1950, which triggered an Acehnese rebellion against Indonesian rule in 1953. Political demonstrations, assassinations, and guerilla warfare would continue off and on again in Aceh for the next half-century, provoking the Indonesian government to alternate between granting concessions to the Acehnese and brutal crackdowns on Acehnese resistance, including widespread "disappearances" of political opponents and other human rights violations committed during the Suharto regime (1968–1998). In the meantime, the discovery of natural gas reserves in Aceh province in the 1970 served only to up the ante, strengthening Indonesia's resolve to control Aceh while simultaneously adding to the grievances of the Acehnese people, who charged that most of the revenues from Aceh's resources were being appropriated by elite groups in Jakarta.

Unfortunately for the Acehnese, tension between Aceh and the central government reached a peak just before the arrival of the Indian Ocean Tsunami. Encouraged by the downfall of Suharto and the independence movement in the former Indonesian province of East Timor, Acehnese activists called for a referendum on Acehnese independence in 1999. When this call was rebuffed by Jakarta, Acehnese guerilla warfare began anew, and once again the Indonesian state's response alternated between concessions (such as finally granting Aceh the right to implement *shari'a* law in 2001) and strongman military tactics. In 2003, just a year before the tsunami, the Indonesia government had begun a repressive counter-insurgency operation which all but cut off Aceh from the outside world, and this conflict was still ongoing when the tsunami struck.

Aceh's political vulnerability at the time of the Indian Ocean Tsunami was compounded by economic weakness. As Bill Nicol (2013) points out, Aceh's ongoing insurgency was inspired, not just by historic grievances, but also by Aceh's chronic modern-era poverty. At the time of the tsunami, 30% of the Acehnese lived below the poverty line, compared to a national average of 17%, and fully 12% of the Acehnese were unemployed. While on paper Aceh was one of Indonesia's wealthier provinces due to oil and gas revenues, in practice, most of this income was siphoned off by the national government. The same was true about Aceh's lucrative illegal logging industry, which since the 1960s has largely been under the control of Indonesia's powerful but corrupt military. Illegal logging was a double drain on Aceh: not only did it extract wealth from the region, it left behind deforested slopes subject to high levels of erosion and soil degradation.

A Crisis Compounded: The Tsunami in Aceh

As a result of chronic economic weakness and political instability, therefore, Aceh was arguably already in a state of crisis even before the Sumatra-Andaman Earthquake struck on December 26, 2004. Given the underlying vulnerability of the population, it is small wonder that nearly three-quarters of all lives lost to the Indian Ocean Tsunami were Indonesian, and most of these Indonesian lives were Acehnese. Aceh's proximity to the earthquake epicenter also magnified the extent of the damage. Since Aceh's west coast was only about 100 kilometers from the quake epicenter, the tsunami reached heights of 20–25 meters along this shoreline and inflicted almost unimaginable destruction. In the fishing town of Meulaboh, an estimated 10,000 were killed by the waves and 80% of the buildings were destroyed. The death toll was even higher in Calang, the regional capital, much of which was built on a vulnerable coastal sandbar. The tsunami all but wiped Calang off the map, transforming a "previously verdant landscape" into "desolate land, littered with the wreckage of buildings and scattered trees" (Umitsu, 2011). Luckily, this part of Aceh was not well inhabited, in part because of recent internal population displacement due to the ongoing conflict between the Indonesian Army and the Free Aceh movement. Paradoxically, then, the ongoing civil war may have saved some lives on Aceh's west coast.

Unfortunately, many of those displaced from the west coast had fled to Banda Aceh, Aceh province's largest city, which also suffered widespread destruction during the Indian Ocean Tsunami. Wave heights were considerably lower in Banda Aceh, between 3–9 meters, in part because a protruding headland to the west of the town blunted the force of the tsunami. Nonetheless, geographic and economic conditions in Banda Aceh ensured that the tsunami wave's impact would be near-catastrophic. Since Banda Aceh was built at the mouth of the Aceh River, nearly the entire city is underlain by river alluvium, which at its highest points only rises about 3 meters above sea level. These lowlands were criss-crossed by canals and waterways, further facilitating the penetration of the tsunami, which would ultimately reach up to 4–5 kilometers into the city's interior. What is more, as Murat Saatcioglu et al. (2007) have pointed out, Banda Aceh's vulnerability was worsened further by the architectural deficiencies of its commercial and residential structures. Most of Banda Aceh's masonry buildings were non-engineered, lightly-reinforced concrete structures that reflected the region's poverty but were a poor match for its seismic realities. Banda Aceh did have some wooden homes as well, which have more resilience against earthquakes, but many of these homes would break into fragments when struck by the incoming tsunami.

Banda Aceh's disaster began at 8am, when earthquake waves rippled through the city of nearly 300,000 people, knocking many of them off their feet. Few in Banda Aceh were worried at the time about a tsunami wave, which is not atypical: it is impossible for a lone observer in a single location to distinguish a moderate, localized earthquake which cannot trigger a tsunami from a more powerful but more distant underseas quake that can. An early warning system of sensor buoys and sirens might have alerted Banda Aceh to its danger, but at the time no such system existed in the Indian Ocean. As a result, many inhabitants of Banda Aceh were out surveying the earthquake damage when tsunami waves began to swamp the city. According to anthropologist Annemarie Samuels (2019), survivors would later describe their experiences with vivid metaphors, describing the tsunami as a huge black wall of water, which sounded like an explosion, moved like a snake, and spun and battered its victims like the blade of a blender. For many, escape from the waters was impossible, as many of Banda Aceh's buildings were only a single story tall and offered no shelter from the incoming flood (Figure 3.2).

Multi-story structures offered more protection from the surging waters, but only if they could withstand the impact of the incoming waves. According to Saatcioglu et al. (2007), many masonry homes and businesses that had suffered structural damage during the quake would later collapse when subjected to the hydrodynamic forces of the tsunami. What is more, Banda Aceh's buildings that managed to weather the quake and withstand the tsunami waves faced a further threat: Banda Aceh's own fishing fleet, much of which had become unmoored

Figure 3.2 An Indonesian village devastated by the 2004 Indian Ocean Tsunami
Source: U.S. Navy photo by Photographer's Mate 2nd Class Philip A. McDaniel.

by the tsunami waves and was now being carried by water currents through the city. It was joined by smaller bits of floating debris as well, including cars, trucks, and floating chunks of buildings, which smashed its way through the city like a flotilla of battering rams. The largest battering ram of all, however, was an oil tanker-sized floating power generator station, which came loose from its dock in Banda Aceh's harbor and was driven 3.5 kilometers inland, leveling entire neighborhoods along the way.

Needless to say, Indonesia's Aceh province suffered almost unimaginable damage as a result of the Indian Ocean Tsunami. According to Bill Nicol (2013), an Australian consultant who aided Indonesia's recovery effort, the retreating tsunami waves left behind a "virtual moonscape" along 800 kilometers of coastline, and, depending on local topography, this moonscape stretched as far as 6 kilometers inland. Some 130,000 houses were destroyed in Aceh, along with 600 government buildings, 2,000 school buildings, and 5,000 health care facilities. In Banda Aceh, Nicol remembered, the jumbled ruins of these buildings formed dunes of debris up to 4 meters high that stretched "as far as the eye could see." Much of Aceh's economic assets were lost to the tsunami as well, including 20,000 fish ponds, 2 million head of cattle, and 60,000 hectares of farmland, which was buried under sediment and debris and rendered infertile by salt infiltration. The World Bank would later estimate that the tsunami disaster inflicted $4.45 billion in damage to Aceh, a sum equivalent to a normal year of Aceh's GDP.

Most importantly, the human cost of the disaster was horrific. Over 167,000 people were dead, to the point that some survivors remembered that Banda Aceh's post-tsunami streets seemed to be paved with corpses. Over 500,000 people were displaced by the tsunami, but these hungry and homeless survivors could expect little assistance in the near future, as the tsunami had also wreaked havoc on Aceh's roads, seaports, and landing strips. What is more, Aceh's plight was slow to come to the attention of the international community, partly as a consequence of Aceh's ongoing civil unrest. As part of their pacification campaign, Samuels (2019) notes, the Indonesian military tried to seal Aceh off from the outside world, meaning that "when the tsunami struck in late December of 2004, hardly any journalists or foreign aid organizations were at the scene." Distracted by cell phone videos from tsunami-stricken Thailand, therefore, the world only gradually came to realize the full extent of Aceh's tragedy.

Aceh's Aftermath: "Setelah Tsunami"

To some degree, it makes sense to describe the 2004 tsunami in Aceh as part of a longer historical process, since the root causes of Aceh's poverty and political turmoil can be traced back to the founding of Indonesia, the Dutch invasion, or even earlier. Nonetheless, as Samuels (2019) points out, Oliver-Smith's contention that disasters are processes rather than events would have found few advocates in post-tsunami Aceh, where the disaster was understood as a sudden and jarring rupture that had turned the world upside-down. According to Samuels, the Indian Ocean Tsunami was a singular event of such profound importance to the Acehnese that it changed their concept of time itself; in the years after 2004, it became common for Aceh's residents to divide the past into *sebelum tsunami*, "before the tsunami", and *setelah tsunami*, "after the tsunami."

The first task *setelah tsunami* was to feed, house, and clothe the survivors in Banda Aceh and elsewhere along the stricken coast. The problem was not one of resources: once the international community realized the depth of Aceh's plight, an outpouring of humanitarian aid was promised to Aceh. As Jasparro and Taylor (2011) point out, modern-day disasters often trigger a phenomenon they call "competitive compassion," where donor nations try to out-bid each other in terms of promised aid, and this process helped to raise $14 billion in relief and rebuilding aid in the tsunami's aftermath. However, getting this aid to those who needed it in Aceh proved to be problematic. An insurgency was still ongoing, and there were some fears that terrorist or Islamist groups masquerading as relief organizations might exploit the situation if given a window into Aceh. Aceh's infrastructure was in shambles: 3,000 kilometers of road had been ruined by the tsunami, most of the region's seaports were inoperable, and most of the region's airfields were off-line, buried under thick marine sediments. Indonesia's swollen and inefficient bureaucracy often proved to be more of a hinderance than a help in getting supplies to the afflicted region. In addition, some feared that foreign aid workers sent to Aceh might be targeted by the sometimes-overzealous religious police who were enforcing Aceh's *shari'a* laws.

Perhaps most importantly, how would the Indonesian military, a state within a state, react to the tsunami? As Bill Nicol (2013) points out, the military could have seized on the tsunami as an opportunity to isolate and root out the Acehnese rebels. What is more, as we saw in Chapter 2, large-scale natural disasters often trigger xenophobic reactions in the surviving populating due to the activation of the behavioral immune system in the presence of disease cues. In this case, however, any biological tendency toward out-group aversion that occurred after the tsunami was apparently resisted. In fact, both the Indonesian Army and the leadership of the Free Aceh insurgency deserve credit for reacting pragmatically to the disaster. On the day after the tsunami, the military announced that foreign aid workers and journalists were now free to enter Aceh, while the Free Aceh fighters declared an immediate cease-fire so that humanitarian aid could reach those who needed it. Since this cease-fire proved to be permanent, Aceh's natural tragedy was not compounded by the man-made tragedy of civil war.

Although peace between the Aceh rebels and the government in Jakarta made rebuilding possible, the task of reconstruction was by no means easy. Before any homes could be rebuilt, the rubble had to be cleared away, and this job quickly turned into a corpse-mining operation in Banda Aceh, where recovery teams extracted as many as 3,000 bodies a day from the debris. Unlike Haiti after 2010, where much of the post-earthquake recovery work was performed by outside contractors, the job of removing the rubble in Indonesia was largely given to the local Acehnese, providing an economic lifeline to out-of-work disaster victims. In fact, many of the new jobs created by the BRR, Indonesia's newly-created rehabilitation and recovery agency, were given to former members of the Free Aceh rebel group in order to co-opt them into the rebuilding program and, by extension, the Indonesian state.

Once the land was cleared, the BRR and overseas NGOs joined forces to rebuild Banda Aceh and other afflicted communities, ultimately constructing over 140,000 homes, 4,000 kilometers of roads, 270 bridges, and close to 2,000 schools and 1,000 governmental buildings. These rebuilding efforts were not without hiccups. Although vast sums were promised by donor nations due to post-disaster "competitive compassion," some of those financial promises were slow to be realized. Many Acehnese criticized the pace of the construction projects, particularly those of the BRR, and some of the money earmarked for disaster relief seems to have been diverted to other projects, such as a new superhighway connecting Banda Aceh to a prospective oil field. Paradoxically, other Acehnese complained that "we can't eat houses," a critique of the government's apparent fixation on rebuilding homes after the disaster rather than rebuilding the broader economy. Still, within four years of the tsunami, much of Aceh was beginning to return to some degree of normalcy.

One thing that did not return to normal, thankfully, was the long-standing hostility between Aceh and the central government in Jakarta. As Nicols points out, the tsunami served as an "effective circuit breaker" for the conflict, bringing a long tradition of civil strife to a welcome end. For the Acehnese, the tsunami seems to have served as a moment of introspection. Some Acehnese felt like the tsunami brought Aceh to the world's attention, validating Aceh's claims of autonomy. However, many more Acehnese, in keeping with Islamic tradition of seeing natural disasters as messages from God, interpreted the disaster as Allah's punishment for the violence of the separatist movement. In any case, the sheer enormity of the reconstruction process made many Achenese more receptive to Jakarta's control, so long as Jakarta provided the post-disaster relief funds that Aceh felt entitled to. For its part, the government in Jakarta was more than willing to trade resources for peace, as epitomized by the government's willingness to incorporate former Free Aceh leaders into the BRR recovery agency.

As I write this, nearly 20 years after the Indian Ocean Tsunami, the peace agreement brokered in Aceh by the Indian Ocean Tsunami remains in place. Aceh today enjoys considerable local autonomy from Jakarta, including the right to elect representatives to a regional legislative body, the Aceh House of Representatives. At the same time, overall Indonesian control over Aceh is no longer in question. As Samuels (2019) points out, this sovereignty is symbolized by the massive tsunami museum that the government constructed in downtown Banda Aceh, a monument to Jakarta's authority and a constant reminder of the Indonesian government's investment in the post-tsunami relief effort. In addition, in keeping with the spirit of pragmatism fostered by the Indian Ocean Tsunami, the four-story museum was designed to serve as a possible shelter-point in the case of future tsunamis. One can only hope it never needs to be used.

Building Vulnerability in Sri Lanka: The Tamil/Sinhalese Conflict

About one hour and 50 minutes after the first waves of the Indian Ocean Tsunami struck Aceh, tsunami crests began to roll up the beaches of Sri Lanka 1,600 kilometers to the west. Although nearly two hours had passed since the Sumatra-Andaman earthquake, few Sri Lankans had been warned of the impending waves, since Sri Lanka, like Indonesia, had no tsunami detection system in place in 2004. This is not because Sri Lanka had never experienced a tsunami before. Far from it: mild tsunamis had occurred in both 1881 and 1883, and Sinhala-Buddhist scriptures preserved the memory of severe historical tsunami that probably occurred between 205–161 BCE. According to the *Mahavamsa* chronicles of Sri

Lanka, a certain King Kelanitissa once ordered a Buddhist monk to be boiled alive in the mistaken belief that he was having an affair with the queen. Offended by the king's impiety, the gods made the sea the instrument of their anger, and ocean waves raged inland and flooded the land. These waves only receded after young Princess Viharamahadevi bravely sacrificed herself to the angry sea by setting out into the waves on a small boat. Viharamahadevi survived her ordeal by tsunami, and would later give birth to Dutugemunu, one of the greatest heroes of Sri Lanka.

For most Sinhalese Sri Lankans at the time of the 2004 tsunami, however, the main lesson remembered from the *Mahavamsa* chronicles was not the tsunami threat, but rather the historical threat posed by the Tamil people of Sri Lanka's north. According to the chronicles, Dutugemunu became the ruler of the Kingdom of Anuradhapura, which was inhabited by the Sinhalese, an Indo-European group of Theravada Buddhists who claimed to have descended from settlers from northern India. During his reign, Dutugemunu's main opponents were the Tamils, Dravidian-speaking Hindus who at that time were expanding into Sri Lanka from southern India. With his victory against the Tamil king Elara, Dutugemunu freed the Sinhalese from Tamil domination. As a result, Dutugemunu is mainly remembered as a champion of the Sinhalese people against the Tamil enemy.

While the events of the *Mahavamsa* chronicles occurred in the distant past, the Sinhalese/Tamil conflict they described probably seemed more relevant than ever in 2004. At the time of the tsunami, an ugly civil war between the Sinhalese majority in Sri Lanka and the Tamil minority was entering its 25th year. The Liberation Tigers of Tamil Eelam (LTTE), better known as the Tamil Tigers, had been formed in 1983 in reaction to the "Black July" pogrom of the same year, during which Sinhalese mobs collaborated with some Sri Lankan officials to kill thousands of Tamils, rape hundreds more, and set fire to thousands of Tamil homes and businesses. The LTTE's goal was to prevent future such outrages by forging an independent state in Sri Lanka's north and east, where Tamils made up the majority of the population, and the tactics they employed towards this end included guerilla warfare, political assassinations, and suicide bombings. By the time of the Indian Ocean Tsunami, the Sri Lankan Civil War had already claimed up to 70,000 lives, including both Sinhalese and Tamils, soldiers and civilians. In addition, 500,000–1,000,000 Sinhalese and Tamils had been displaced by the fighting. At the time of the tsunami, Norwegian mediators had secured a cease-fire in the conflict, but tensions remained. What is more, a large swath of northeastern Sri Lanka remained under effective Tamil Tiger control, including large parts of the Eastern Province, which would be badly damaged by the Indian Ocean Tsunami.

The ongoing civil war was not the only source of vulnerability faced by Sri Lankans in the lead-up to the tsunami. In contrast to Aceh, where the tsunami struck both rural and urban populations, the portion of Sri Lanka which was struck by the tsunami was predominantly rural, dotted with small fishing villages, agricultural towns, and a few tourist enclaves, including the small city of Galle. As Piers Blaikie points out, much of this shoreline was occupied by the poorest of Sri Lanka's poor, including Tamil and Muslim farmers, marginal people who had been displaced to marginal coastal land by years of discrimination and conflict as well as deliberate government-sponsored Sinhalese colonization schemes dating back to the 1950s till the 1970s (Blaikie and Lund, 2010). Many other inhabitants of the region were fishermen and their families, who often occupied flimsy houses made of wood and straw with mud walls that were close to the beach, their nets, and their boats. With the exception of the "Fort" neighborhood in Galle, which boasts thick defensive ramparts of coral and granite stone that are a legacy of Portuguese and Dutch colonial occupation, this coastline lacked any significant seawalls that might have curbed an incoming tsunami.

Geographic factors contributed to Sri Lanka's tsunami vulnerability as well. The southern and eastern coastal plain of Sri Lanka is extremely flat, other than occasional sand dunes, and for the most part offered little resistance to the incoming tsunami waves. What little resistance the coastline did enjoy was by virtue of its coastal mangrove forests, but large portions of this forest had been destroyed by 2004, lost to tourist development, firewood collection, and to the construction of commercial fishponds in the tidal flats.

The "First Wave", the "Golden Wave," and the "Third Wave" in Sri Lanka

As is often the case when tsunamis come ashore, the first part of the oscillating tsunami wave to strike Sri Lanka was the trough, so the tsunami signaled its imminent arrival by sucking water off the shoreline to a distance of nearly a kilometer. However, few in Sri Lanka recognized the signs. Rather than running inland, many fishermen ran to the beaches to save their boats and nets from being pulled out to sea. In other places, children flocked to the waterfront to see the spectacle of the exposed shoreline and to collect the flopping fish left behind by the receding waters. Still other Sri Lankans began to drive stakes into the newly-exposed shoreline, laying claim to ownership of what they thought was newly-emerged land. Then, with clockwork predictability, the tsunami waves flooded back inland, cresting to heights of approximately 5–10 meters depending on local topography and reaching a kilometer or more inland. Overall, 65–70% of Sri Lanka's coastline was affected, with the worst damage and wave infiltration occurring on the eastern coast, which runs north-south and thus was broadsided by the tsunami's westward-flowing waves. Damage was more localized along the southern coast, which received a more glancing blow and which was protected by some headlands. Nonetheless, some parts of the south suffered significant destruction due to misfortunes in local topography, such as V-shaped bays, which concentrated the energy of the tsunami waves.

Witnesses to the tsunami in Sri Lanka remembered that the incoming waves made the noise of an incoming storm and described the wave crests as "balls of water," high enough to graze the treetops. Since the tsunami first scoured out the tidal zone before flooding back in, the incoming waters were brown or orange, thick and sticky with sediment. One survivor later told researchers that, as she clung for life to an overhead ceiling fan, detritus rushed past by outside, "mattresses, a TV, fridges, coconuts, chickens, all in a froth." The same witness also complained that the water itself seemed poisonous, causing her skin to itch and bubble up. Researchers would later find that the tsunami water had become contaminated with hundreds of different types of bacteria as it swept through unclean city streets and washed through sewers and cesspits, and inhalation of this water lead to a characteristic respiratory infection in many survivors that rescue doctors would dub "tsunami lung."

As a result of the region's flat topography and relative lack of multistory structures, the death toll from the tsunami in Sri Lanka was enormous, second only to that of Aceh in Indonesia. In all, 35,322 Sri Lankans died and over 100,000 homes were fully or partially destroyed, leaving more than half a million people homeless. Property damage was estimated at $1 billion, an amount equal to 4.5% of Sri Lanka's GDP at the time. As Mulligan and Shaw (2011) point out, the impact of the tsunami was felt most acutely by the "poorest and must vulnerable population groups: fishing communities, shanty dwellers, and those living in the northern and eastern provinces, which prior to the tsunami had been ravaged by two decades of civil war." As was the case throughout the Indian Ocean, the tsunami killed approximately four times as many women as men, in large part because men were more likely to know to how to swim, had more experience climbing trees, and had more upper body strength to resist being swept away by the surging currents.

In terms of geography, the worst-affected provinces were Ampara in the east, where tsunami waves killed over 10,500 people, and Galle in the south, whose high death toll of over 7,000 reflects that city's relatively high population density as much as the height of the waves. Many of the Galle victims died during what has been described as the worst train disaster in history, the derailment of the "Queen of the Sea," which on the day of the tsunami was heading south on a coastal rail line and was packed full of travelers who had spent Christmas in the capital city of Colombo. The train stopped in the village of Seenigama when it was struck by the first wave, and its apparent stability made it a magnet for the local population, who flocked aboard the halted train for shelter. Unfortunately, ten minutes after the first wave, a second and stronger tsunami wave that may have reached 9 meters in height overtopped and then toppled the train. At least

1,700 people died in this tragedy, drowned in the overcrowded train cars or crushed by the train itself as it was pushed inland by the flood.

In part because the tsunami struck the tourist center of Galle, Sri Lanka's tsunami tragedy quickly came to the attention of the world community, and humanitarian aid almost immediately began to flow into Sri Lanka's stricken communities. Due to the weakness of the Sri Lankan state, much of this aid was provided by NGOs; by January of 2005, in fact, no fewer than 890 different international relief agencies were carrying out operations in Sri Lanka. Sri Lankans called this sudden influx of resources the "second wave" or the "golden wave," and it inspired mixed opinions. While some in Sri Lanka were very thankful for the foreign assistance, others did not like the feeling of dependency and indebtedness it entailed, and still others believed that resources were being taken by those who did not deserve them. According to Michele Ruth Gamburd (2014), many Sri Lankas watched in outrage as "people who merited no aid" and lived well outside the disaster zone received "gas cookers, sewing machines, bicycles, fishing boats, and new houses – all significant indices of prestige." Thus, Gamburd continues, "the 'golden wave' functioned as summarizing symbol ... [of] people's concerns about inequity, immorality, lack of accountability, and artificially inflated social statues."

What is more, much like the 1855 Ansei Edo earthquake discussed in Chapter 2, Sri Lanka's "golden wave" threatened to undermine existing social hierarchies. The rich, who had the most property to lose, suffered the worst economic losses during the tsunami, while the destitute received substantial material aid and benefitted from wage inflation during the post-tsunami reconstruction program. Indeed, many Sri Lankans believed that generous humanitarian aid actually slowed down reconstruction by keeping some workers out of the job market, since they were being fed, housed, and otherwise supported by post-disaster charities. Others objected to the fact that those who lost houses were given a lump-sum payment for reconstruction, regardless of the actual value of the destroyed house. In sum, as one of Gamburd's interviewees complained, following the golden wave, "people don't remember who they are and what their status is."

The post-disaster recovery efforts of the Sri Lankan government were criticized even more harshly than those of the foreign NGOs. As Gamburd explains, Sri Lankan politics has long been dominated by a patronage system, "in which leaders exchange votes for delivering projects, favors, and funds to members of their electorate." In one egregious example of the corrupt administration of tsunami relief funds, the small town of Hambantota, which was the political base of the sitting Sri Lankan president, received post-disaster housing assistance equal to five times the actual value of the town's damaged homes. In general, the mostly-Sinhalese south received far more than its fair share of reconstruction funding, while the predominantly Tamil east was seriously short-changed. While some aspects of the "golden wave" may have undermined existing socioeconomic hierarchies, therefore, Kalinga Tudor Silva (2010) has concluded that overall "the tsunami response clearly reinforced the existing political and ethnic prejudices of the state."

The state's favoritism toward the Sinhalese also seemed to be on display in its post-tsunami edicts on coastal "buffer zones." The starting line of these zones was initially set at 200 meters from the sea in the Tamil east of Sri Lanka but at only 100 meters in the south and west, where the Sinhalese predominated, a difference that many Sri Lankas ascribed to state favoritism toward the Sinhalese. However, some critics have interpreted this decree differently, ascribing it not to ethnic prejudice, but neoliberal capitalism. Noting that this exclusion zone decree forbade the rebuilding of houses within the buffer zone, but allowed commercial development, Naomi Klein claimed in 2007 that its real purpose was to dispossess local fishermen of their land and open it up to exploitation by foreign investors (Klein, 2008). Klein calls this phenomenon "disaster capitalism" and presents it as a specific application of her broader "shock doctrine" thesis, the notion that elite groups often exploit the aftermath of disasters to push for fundamental re-ordering of society, usually to the detriment of the marginalized poor.

In his own work on the fishing communities in post-tsunami Sri Lanka, however, Arvid van Dam (2015) presents a more nuanced view of the recovery process. While some fishermen were effectively excluded from the sea by these new rules and were forced to sell their land to capitalist developers, other fishermen were tacitly allowed to re-occupy their land if it was not suitable for commercial development or if they agreed to absolve the government of any responsibility for their protection. What is more, van Dam notes, many fishermen are finding creative ways to make money from tourism, for example, by selling their catch directly to hotels, leading fishing safaris for visiting vacationers, or even setting up small-scale guest houses to accommodate visitors. In any case, the local fishing industry was in crisis even before the tsunami, due to rising fuel costs and declining catches caused by overfishing and foreign competition. The overall picture van Dam presents therefore is not dualistic opposition between local victims and foreign capitalists or between fishing and tourism, but rather creative negotiation between stakeholders, in some cases, for their mutual benefit.

By the time that van Dam was doing his fieldwork in Sri Lanka, these fishermen had already survived what many Sri Lankans called the "third wave" – the abrupt withdrawal of many NGOs in late 2005. While some NGOs left after completing their self-appointed projects, others evacuated suddenly due to the outbreak of renewed civil conflict in Sri Lanka (which we will discuss below) or the diversion of their resources to a new disaster flashpoint, the destructive Kashmiri Earthquake of October 2005. As a result of this third wave, many housing projects were left unfinished and many Sri Lankans who had become accustomed to foreign financial support were left to their own devices. Thousands of Sri Lankans were therefore still trapped in temporary camps a year after the Indian Ocean Tsunami, where they suffered from "overcrowding, poor sanitation and lack of drinking water, and high levels of alcoholism, domestic violence, and sexual abuse" (Milligan and Shaw 2011).

The third wave was a jarring reality check for many Sri Lankans, who were inspired by the "Build Back Better" slogan of the tsunami recovery effort, a promise that ultimately went mostly unfulfilled. But perhaps Sri Lankans should not have been surprised. As Silva (2010) points out, the international aid industry typically "moves fast where disaster strikes, provides emergency relief, and moves on to the next disaster site without necessarily completing the tasks undertaken." In addition, Silva continues, the work of NGOs usually "does not build on any existing processes or leave behind any viable entities that add to the resilience and coping mechanisms of the affected populations." While the international relief effort in Sri Lanka almost certainly saved some lives, therefore, it is highly debatable whether the vast sums spent during the "golden wave" – the equivalent to about $7,100 for every person in the disaster zone – have had any lasting positive effect on Sri Lanka. Indeed, according to Sarah Khasalamwa (2010), Sri Lanka's poverty rate actually climbed from 64% in 2004 (before the tsunami) to 80% three years later.

Renewed Civil Conflict in Sri Lanka: the Failure of P-TOMS

As was the case in Indonesia, the Indian Ocean Tsunami struck Sri Lanka in the midst of a civil war, and in the immediate aftermath there was some hope that the national tragedy might bring the two parties closer together. At the time of the tsunami, physical control over the island was divided between the Sri Lankan Army and the LTTE, which created both a problem and an opportunity. The problem was simple: how can a nation distribute relief aid in a portion of the country that is effectively under the power of a rebel faction? But this problem, in turn, created an opportunity. If a framework could have been agreed upon for cooperative action between Sri Lankan authorities and the LTTE, it could have served as a blueprint for lasting peace between the two groups.

In May of 2005, a group of international donors came up with a framework that might have fit the bill – the Post-Tsunami Operational Management Structure, or P-TOMS. Initially, the government of Sri Lanka signed on to this agreement, as did representatives of the LTTE and the Muslim minority party in Sri Lanka. However, Sinhalese nationalist hardliners were vehemently opposed to this arrangement. In addition to compromising Sri Lanka's international sovereignty by granting too much authority to the World Bank, Sinhalese nationalists argued that P-TOMS accorded quasi-governmental status and *de facto* political legitimacy to the LTTE, which in their minds was nothing but a terrorist organization.

Opposition to P-TOMS became a rallying point of Sinhalese nationalists in the elections of late 2005 and contributed to the narrow victory of Mahinda Rajapakse of the Sri Lanka Freedom Party. As a result, P-TOMS was shelved, and a window of opportunity was closed.

With Rajapakse's election, Sri Lanka quickly slid back into active civil war. The government ceased providing tsunami relief aid to areas under control by the LTTE and moved to sideline the Norwegian team monitoring the cease-fire. The LTTE in turn resumed its bombing campaign and shut off the spigots of the Mavil Aru reservoir, which was the main water source for 15,000 people in government-controlled Sri Lanka. The government responded with a new military offensive against the LTTE in 2006, even though the cease-fire agreement was still formally in effect. The war dragged on for three more years, led to tens of thousands more deaths, and displaced 200,000 civilians, triggering a new humanitarian disaster for Sri Lanka. In the end, Rajapakse and his government claimed victory over the LTTE, though this accomplishment was tainted by widespread accusations of war crimes and human rights violations committed by government troops during the fighting. In any case, as Mulligan and Shaw (2011) point out, governmental victory in the civil war came at a great financial and human cost and has "done nothing to reduce the ethnic divisions in Sri Lanka," which remains "badly divided."

The Indian Ocean Tsunami, therefore, interacted with ongoing civil conflicts in Indonesia and Sri Lanka in sharply different ways. While the tsunami helped convince Jakarta and the Free Aceh rebels to sign a peace agreement, the tsunami may have exacerbated the conflict in Sri Lanka by highlighting the problems of divided sovereignty on the island. These disparate outcomes, in turn, underscore the importance of considering the political, cultural, and socioeconomic backgrounds of societies impacted by similar disasters. In the final analysis, the fault lines dividing Sri Lankan society simply ran deeper than those separating Aceh from Indonesia – the Sri Lankan conflict had older historical roots, the religious divide was starker, and the ethnic differences between Tamils and Sinhalese were more pronounced. As a result, while the Indian Ocean Tsunami may have contributed to the end of the Aceh-Indonesian conflict, millions of tons of cold water from the Indian Ocean proved insufficient to shock Sri Lankan society into a negotiated peace.

The Tōhoku Tsunami and Nuclear Disaster, 2011

In 2006, just two years after the Indian Ocean Tsunami, Japanese tsunami experts Kenji and Tomoko Yamazaki carried out field research in Seenigama, Sri Lanka, the site of the "Queen of the Sea" train derailment. They interviewed families, collected demographic data on both the living and the dead, and mapped damage to houses and inundation levels. In addition, they offered some condescending words of advice to Sri

Lankans. One of Seenigama's problems, they suggested, was the lack of a viable tsunami culture; although Sri Lanka had been struck by tsunamis before, "the people of Sri Lanka did not pass on their sad experiences … to their descendants," and as a result, many Sri Lankans died on December 26, 2004 because they lacked "adequate knowledge of tsunamis" (Yamazaki and Yamazaki, 2011). Yamazaki and Yamazaki also contrasted Seenigama's failures to the success of Taro village in Japan's tsunami-prone Sanriku Coast. After being leveled by a 1933 tsunami, Taro had developed a robust "tsunami prevention subculture," which included early warning systems, evacuation plans, public education campaigns, and above all the construction of an elaborate seawall. As a result, the authors note, Taro has been struck by four different tsunamis since 1933, but "there has been no loss of human life … [as] Taro has found ways to mitigate tsunami disasters."

Unfortunately, Yamazaki and Yamazaki's hubristic praise of Taro's "tsunami prevention subculture" must have provoked the gods: just four years later, Nemesis arrived in Taro in the form of a 15-meter tsunami. Taro's pride, the 10-meter seawall that was popularly dubbed the "Great Wall of Japan," was quickly overtopped, and Taro itself was razed down to its foundations. When rescuers arrived to survey the damage, they had to locate the town via GPS, since the town's site had been transformed into a "sandy, barren plain." First responders found that only two structures remained standing: a waterfront building so battered as to be unrecognizable, and a hotel with three gutted lower floors and a car tire lodged between the metal rails of a fourth-floor balcony. Almost nothing else remained.

Taro's much-praised tsunami prevention subculture, therefore, did not prevent the destruction of the town. It did, however, save the lives of most of Taro's inhabitants. As luck would have it, Taro had held its annual evacuation drill just a week before, so most citizens were aware of what to do in case of a tsunami. Some of Taro's 5,000 inhabitants set out for designated shelter sites in the hills upon feeling the tremors of the quake, while others did the same when the town's warning sirens began to blare out the alarm, and Taro's wide streets and well-marked evacuation routes facilitated their escape. Most of those who remained watched the sea closely and headed inland as soon as the first swells of the tsunami approached the shore. While the tsunami did claim about 200 victims in Taro, many of them died, not because they were unaware of the tsunami risk or were ignorant of the evacuation procedures, but because they were unable to evacuate quickly due to age or infirmity. Others perished in Taro because they disregarded what they had been taught and climbed atop Taro's seawalls, hoping the incoming tsunami would be harmless entertainment rather than the catastrophe it proved to be.

The quake that spawned Taro's tsunami was the Tōhoku Earthquake of 2011, named after the Tōhoku region of Japan, the cold and relatively thinly populated northern quarter of Japan's largest island, Honshu. Some sources call the same event the Sanriku Earthquake, after the rocky, indented coastline of northern Tōhoku that runs from the Oshika Peninsula to Honshu's northern tip. By whatever name, this quake was extremely powerful, in fact unexpectedly so. Scientists had previously believed that even though the boundary between the Pacific oceanic plate and the Okhotsk continental plate was a subduction zone, the specific characteristics of this fault would prevent any quake from exceeding 8 in moment magnitude. But the Tōhoku Earthquake clocked in at 9.0–9.1 moment magnitude, 10 times what was anticipated. In a matter of seconds, the entire island of Honshu shifted 2.4 meters to the east, and the Pacific plate slipped 50 meters relative to the Okhotsk plate, displacing an enormous amount of seawater. The Tōhoku Earthquake was even more powerful than the Jōgan quake that gave tsunamis their name back in 869; while the two earthquakes struck the same region and flooded the same coasts, it is now estimated that the Jōgan quake only reached 8.4 in moment magnitude.

Vulnerability and Vigilance along Japan's Tsunami Coast

As might be expected of the part of the world that coined the term tsunami, the Tōhoku coast is exceptionally prone to these events. In addition to lying a mere 100 or less kilometers from a subduction zone of the fast-moving Pacific plate, the local geography is highly favorable to flooding. In the Sanriku Coast area, which includes the town of Taro, the coastline is saw-toothed and jagged, interrupted by frequent narrow inlets locally known as *ria*. These *ria* are a boon for fishermen, because they create sheltered harbors, but they also funnel and concentrate the power of tsunami waves, magnifying their effect. Farther south, in the Sendai plain below the Oshika Peninsula, the coastline becomes much smoother, but is also much flatter, consisting mainly of alluvium deposited by the Nanakita, Natori, and Abukuma rivers. There is little to prevent a tsunami from penetrating deep into the interior at this point, though thankfully much of this plain consists of rice paddies and other farmland rather than built-up neighborhoods. Farther south still, in Fukushima, the Tōhoku coastline becomes a mix of flat lands that are prone to tsunami flooding and elevated coastal hills that are somewhat more protected. The Fukushima Daiichi nuclear power plant was located on just such a hill, but unfortunately, the builders lowered and flattened the site considerably before constructing the plant, negating the anti-tsunami advantages of the location.

Why, you might ask, did the Japanese construct a nuclear plant along this tsunami-prone stretch of coastline in the first place? Answering that question requires us to step back and discuss both Japan's energy needs and the economic history of the Tōhoku region. As an industrialized nation, Japan has an enormous appetite for energy, but its domestic supplies of coal and petroleum are meager. Japan was therefore an eager adopter of nuclear power – despite the atomic trauma of Hiroshima and Nagasaki – and in 2011 boasted 56 nuclear plants that together produced almost 29% of the nation's electricity. Japan also turned to nuclear energy for strategic reasons. Because Japan agreed to disarmament after its defeat in World War II, it was important for Japan to project strength in some other way for the purpose of deterrence, and a robust nuclear energy sector was a signal to possible rivals on the geopolitical stage that Japan, if threatened, had the capacity to build atomic weapons.

But why build a plant in Tōhoku, especially given that almost all of Fukushima Daiichi's electricity was being consumed in distant Tokyo? One reason was the "not in my backyard" phenomenon – while the Tokyo megalopolis was happy to illuminate its skyline using nuclear energy, it preferred to keep its reactors at a safe distance. Local authorities in Tōhoku, in turn, were willing to host these power plants because they brought jobs and investment to a region that had been an economic backwater since the Shogunate period and was becoming more so over time with the steady decline of Japan's fishing industry. What is more, the region was hemorrhaging young people – Tōhoku's share of Japan's population had dropped to only 7% by 2011, where it had stood at 11% in 1950, and it was hoped that nuclear investment in the region might staunch this flow.

Fair enough, but why construct the plant on the edge of a seashore known for tsunamis? Economics factored into that decision as well. Nuclear power plants are essentially steam engines, which transform the heat of nuclear fission into electrical power by using superheated water to spin turbines. Before that water can be used again, it must be cooled, and in most nuclear plants this task is performed by the massive conical cooling towers whose tapered profile is synonymous with nuclear power. However, a nuclear reactor built along the Tōhoku coastline could scrap the expensive cooling towers and use the limitless water of the Pacific Ocean to cool the reactors instead. Of course, this decision entailed risks, but the Tokyo Electric Power Company (TEPCO) was confident they could be managed. Too confident, as it turned out.

Nonetheless, while Tōhoku's tsunami-prone geography and economic weakness increased its vulnerability to tsunamis, the region was by no means helpless against tsunami hazards. As we saw in the Taro example above, nowhere in the world understands the threat of tsunamis better, and few places match Tōhoku's level of preparation. Fully 40% of the Tōhoku coast was protected from tsunamis by some combination of breakwaters, floodgates, and seawalls, some of them as imposing as Taro's 10-meter walls. These man-made defenses were bolstered by natural ones, including coastal forests deliberately preserved or planted to blunt a tsunami's force. Early warning systems connected to offshore sensors were designed to trigger automatically in the case of quakes, and many local towns had recently established direct satellite links to get real-time earthquake and tsunami information from the Japan Meteorological Agency. All coastal towns had designated tsunami shelters, the routes to which were marked on widely published hazard maps and street-level signage. Taro had even invested in solar-power lighting that could illuminate tsunami evacuation routes at night in case of an earthquake-induced power outage.

Equally important, though harder to quantify, is Tōhoku's long experience with tsunamis. The local population, which included survivors of previous tsunami events, had a culture of tsunamis that Sri Lanka and Aceh lacked. That being said, Tōhoku's city and regional governments were able to translate that culture into effective anti-tsunami practices in large part due to Tōhoku's relative affluence. Tōhoku is poor only by Japanese standards: with a per capita GDP of $23,950 in 2011, Tōhoku enjoyed nearly five times the per-person wealth of Sri Lanka in 2004 and about twenty times that of Indonesia. Thus, it is debatable whether the Tōhoku coast's "tsunami prevention subculture" was really the crucial factor in saving lives. Without the necessary resources to invest in expensive seawalls and warning systems, resources that Aceh and Sri Lanka did not enjoy, Japan's culture of tsunamis might have been of limited use.

As for the Fukushima Daiichi nuclear power station, it too was far from helpless in the face of tsunami waves, though not those as powerful as the Tōhoku Tsunami. As protection against earthquakes, the plant was anchored on solid bedrock, and it was shielded from tsunamis by two breakwaters and a 5.6-meter seawall, which when it was constructed in the late 1960s was thought to be sufficient to withstand the worst tsunamis the local fault lines interactions could produce. At the time of the tsunami, the TEPCO was belatedly reconsidering this seawall height in response to a 2001 paper by K. Minoura et al. (2001), which had revised the likely severity of the Jōgan Tsunami and Earthquake substantially upwards and which speculated that a similar event was overdue. Still, Fukushima Daiichi had other safeguards in place in case the seawalls were overcome. By the time a tsunami arrived, the plant would already have been in shutdown mode, which was automatically triggered by large quakes. Even after being shut down, the reactor cores needed to be cooled to prevent a meltdown, but TEPCO had designed multiple systems to make sure cool water could still be pumped into the reactors, including on-site diesel generators and emergency battery banks.

The "Compound Disaster": 3/11

Within 10–30 minutes of the earthquake, depending on the proximity of coastline to the epicenter, tsunami waves began to overwhelm Japan's tsunami defenses throughout Tōhoku and the Sanriku Coast. According to Richard Lloyd Parry (2017), who interviewed multiple tsunami survivors, the incoming tsunami was a strange blend of land and sea. It smelled like "brine, mud, and seaweed," survivors remembered, but the sounds it made – "crunch and squeal of wood and concrete, metal and tile" and the "twisting and tearing of timber" – were more appropriate to an avalanche than an ocean wave. As was the case in Aceh during the Indian Ocean Tsunami, the waves carried hundreds of boats inland, including one fishing boat weighting 330 tons that was eventually deposited in a field 600 meters from the shore. Another Japanese vessel, the *Ryou-Un Maru* (in English, "Fishing Luck") was washed from its mooring and by the tsunami and spent the next 13 months wandering the Pacific Ocean as a ghost ship. Its luck finally ran out in April of 2012, when it was spotted off the coast of British Columbia and was sunk by the U.S. Coast Guard after being declared a hazard to navigation.

The severity of the tsunami would depend on local topographic factors as well as the efficiency of the region's various seawalls and breakwaters, but almost the entire Tōhoku coast was submerged. In terms of vertical extent, the tsunami may have topped 40 meters in some enclosed *ria* of the Sanriku Coast, leading to almost total devastation. The tsunami reached its greatest horizontal extent in the Sendai plain north of Fukushima Daiichi and south of Sanriku, where it penetrated 4–5 kilometers from the coast. While Tōhoku's seawalls may have saved some lives by slowing the onslaught of the tsunami by a few minutes, they generally proved entirely inadequate to prevent damage from a tsunami on this scale. Worse yet, the seawalls arguably worsened the disaster in some instances. For those at street level, the seawalls blocked the view of the ocean from the town, eliminating a visual cue of a coming tsunami. In some cases the seawalls prevented the tsunami from returning to sea, leaving the town steeped in polluted water long after the tsunami was over (Figure 3.3). What is more, Kenji Yamazaki argues that Tōhoku's seawalls bred complacency and might have discouraged some Japanese from evacuating due to a misplaced faith in the walls' protection (Yamazaki and Yamazaki, 2011).

Because of the efficiency of the warning system and Japan's robust culture of disaster preparedness, many Japanese had already arrived at designated evacuation sites the time the tsunami waves arrived at their towns. Unfortunately, because of the sheer scale of the disaster, a number of these evacuation sites were themselves overwhelmed by the tsunami. Lucy Birmingham and David McNeill's (2012) book on the Tōhoku Tsunami recounts a particularly harrowing story that unfolded in an elementary school in Higashi-Matsushima, which lies about 3 kilometers from the seashore in the well-populated northern stretch of the Sendai coastal plain. Since the gymnasium of the school was a designated tsunami evacuation zone, it was packed with about 100 children and 100 adults, half of whom were elderly evacuees from a nearby nursing home, within 25 minutes of the earthquake.

At around 3:10pm the occupants of the gym watched with horror as a thundering wall of water, "like a black monster, matted with debris,"

Figure 3.3 Waves of the 2012 Tōhoku Tsunami overtopping the seawall in Miyako
Source: The Headquarters for Earthquake Research Promotion, Japan. https://www.static.jishin.go.jp/resource/figure/figure001015.jpg

plowed through the school's parking lot, picking up or tossing aside the cars as it came. One car smashed the windows of the gym's door, and water began to pour through, quickly turning the gymnasium into a "rising, sucking pool," like a laundry machine spinning as it fills with water. Unable to escape, the evacuees had to tread water to survive, or clung to floating gym mats or furniture while the freezing ocean water drained them of body heat. When the water finally receded later that evening, only 50 people were left alive, and they shivered together in the growing darkness, huddling together for warmth and waiting for morning. Many Japanese who survived the tsunami itself would die of exposure during the bitter cold of the first night after the tsunami, during which temperatures dropped below freezing.

In total, 1,120 people died in Higashi-Matsushima, where the tsunami may have topped 10 meters, easily overtopping the town's 6-meter seawall. Many of these deaths occurred in designated tsunami shelters, and Higashi-Matsushima was far from unique in this regard. In neighboring Ishinomaki, where the tsunami swamped almost half the city, nearly a third of the designated evacuation sites were struck by the tsunami. All told, 3,819 people died in Ishinomaki, which accounted for a full one-fifth of the tsunami's death toll. The highest death tolls per capita, however, occurred in the towns of Otsuchi and Rikuzentakata on the Sanriku Coast, where about 10% of each town's population perished. In Rikuzentakata, a commercial fishing village located in a narrow *ria* inlet, the tsunami topped

17.5 meters, nearly three times the height of the town's seawall. Given the scale of the tsunami, it is a tribute to the efficiency of Rikuzentakata's warning system and evacuation procedures that the death toll was not even worse.

If the severity of the Tōhoku Tsunami was measured purely by the number of casualties it inflicted, it would have qualified as a significant event, though its 19,749 confirmed deaths represent only a small fraction of the casualties inflicted by the 2004 Indian Ocean Tsunami or the Haitian Earthquake of the year before. What differentiated the Tōhoku Tsunami from those other events was the fact that it was a "compound disaster"– the tsunami not only unleashed a deadly flood of water, but also a potentially fatal plume of radioactive particles, released during the meltdown and partial explosion of the stricken Fukushima Daiichi nuclear plant. This combination of tsunami and nuclear disaster was dubbed 3/11 by the Japanese, not just because it occurred on the 11th of March, but also because the Japanese regarded it as an existential threat on a par with America's own 9/11 disaster.

When the Tōhoku tsunami swept into the Fukushima Daiichi nuclear plant, three of its reactors were down for maintenance, and the other three were in cool-down mode, which had been triggered automatically when the plant's instruments detected an earthquake. So far, so good. But nuclear reactors cannot be turned on and off like a light switch; even after the control rods were inserted into the fuel pile to stop the nuclear reaction, radioactive decay processes that were already ongoing continued and heat built up within the reactor. Knowing that this would happen, TEPCO had designed emergency cooling systems into the reactor design. Unfortunately, these systems required electricity to operate, which could be provided either from Japan's power grid, diesel generators at the site, or in a last resort from the plant's emergency battery arrays, which were thought to have enough power to keep the plant going until outside assistance arrived. In the aftermath of the tsunami, however, all three power sources failed. The tsunami and earthquake had severed the electrical lines to the national grid, the tsunami waves had damaged all six of the site's emergency generators, and several of the plant's battery rooms were flooded with seawater. Nor could emergency generators be brought in quickly from off-site; as James Mahaffey (2014) points out in his autopsy of the Fukushima nuclear disaster, all roads leading to the reactor complex "were either completely washed away, blocked by collapsed buildings, or jammed by fleeing people." At this point, a melt-down was inevitable, as Fukushima Daiichi's engineers were quite literally powerless to stop it.

Once the cores of the three active reactors began to melt down, radioactive isotopes such as iodine-131, strontium-90, and cesium-137 began to build up in the containment buildings around the reactors. So too did hydrogen gas, created by the oxidization of the exposed zirconium core supports in the nuclear pile. In essence, Fukushima Daiichi's containment buildings were becoming "dirty bombs," explosives packed with radioactive materials. Between March 13th and March 15th, these dirty bombs detonated repeatedly, releasing columns of nuclear fallout that the prevailing winds blew to the northwest of Fukushima Daiichi. To try to forestall further explosions, TEPCO was forced to vent some gas from the containment buildings, releasing still more radioactive materials in the process.

In response, the Japanese government declared a mandatory evacuation zone around the stricken plant, which was originally set at only 3 kilometers but was later extended to 10, then 20, then 30 kilometers as conditions as Fukushima Daiichi continued to deteriorate. Once the radioactive plume was detected, this evacuation zone was extended still further. As a result, an estimated 340,000 residents of Fukushima province became "nuclear nomads," displaced from their homes and forced to move repeatedly (up to six times in some cases) due to the escalating and expanding evacuation orders. The stress of these evacuations would later be blamed for contributing to the premature deaths of 1,000 elderly residents of Tōhoku. In the meantime, hostility to the government grew in the Fukushima region. Some residents believed that the government was trying to cover up the disaster rather than address it. Others worried that evacuation orders were being issued too late, only after elevated radiation levels had already been detected. Many in Tōhoku probably agreed with the complaint of Fukushima farmer Takao Takahashi, who told a reporter for the *New York Times* on April 5 that "we have been sacrificed so that Tokyo can enjoy bright lights."

Decontaminating and Rebuilding after 3/11

When the tsunami waves receded from the Tōhoku coast, they left behind a world that was almost unrecognizable. Shaken survivors had trouble getting their bearings in what was left of their towns: familiar landmarks had disappeared, and whole neighborhoods had seemingly been vaporized, as if an atomic bomb had wiped them away, or else were now overlain by thick deposits of debris. The amount of rubble left behind by the tsunami in Fukushima, the Sendai Plain, and the Sanriku Coast was estimated at 24–25 million tons. In the town of Kamaya, Parry (2017) tells us, survivors described this wreckage as including "large sections of houses that had been picked up and then dropped by the wave, cars and vans, upended and crushed, and the smallest household items: shoes, sodden garments, cooking pots, teapots, spoons." The tsunami's debris was glued together by stinking black mud, though at Kamaya it also smelled oddly of pine sap, as the tsunami had ripped up the hardwood forest planted to protect the village from tsunamis after 1933 and driven the uprooted trees through the town. In all, the Tōhoku Tsunami inflicted an estimated $235 billion in damage, which if true would make it the costliest natural disaster in history, and killed about 20,000 people, of whom 54% were aged 65 or older.

For many of the survivors, the first task of the living was to recover the memory of the dead. As Fuyubi Nakamura (2012) points out, members of Japanese families have an obligation to pray for departed ancestors, which is typically performed in household shrines that include pictures and/or memorabilia of the dead. However, the tsunami severed this link between the living and the dead, destroying these shrines along with the houses that contained them. As a result, dedicated "memory search" teams spent months after the disaster sifting the tsunami debris for lost and damaged photos and mementos. Indeed, the Tōhoku Tsunami may be the first disaster in which photo retouching software played an essential role in the post-disaster recovery effort.

Unfortunately, these necessary tasks could not be carried out in the nuclear exclusion zone, which no doubt added significantly to the trauma of the nuclear nomads. Before they could return home, their land had to be decontaminated, a time-consuming process that in most places involved scooping up, bagging, and disposing of the top few centimeters of contaminated soil. In some cases, evacuees had to wait a decade before being allowed back home – evacuation orders were not lifted for the village of Okuma until 2019, and Futaba, the town where Fukushima Daiichi is located, was not re-opened until 2022. In the meantime, those displaced by the nuclear disaster suffered badly from psychological stress during their long stay in temporary housing. According to the Fukushima Health Management Survey, which was established after the nuclear disaster to watch over the well-being of the nuclear nomads, Fukushima evacuees suffer from high rates of hypertension, diabetes, and liver dysfunction, the latter due in large part to alcohol abuse. Not surprisingly, Fukushima disaster evacuees also suffer from high rates of depression and suicide.

What the Fukushima Health Management Survey has *not* documented, at least to a significant degree, is any negative health impacts from radiation exposure among the Fukushima evacuees. While the 2011 Fukushima nuclear disaster is often compared to the 1986 meltdown at Chernobyl, in actual fact, Fukushima released only a small fraction of Chernobyl's radiation and Fukushima's radiation was spread over a far smaller area. What is more, while 31 people are known to have died of exposure to Chernobyl's radiation, the number of deaths directly attributable to radiation from the Fukushima nuclear accident still stands at zero as of 2023. This number may rise in the future, as cancers caused by radiation exposure can take years to develop, but based on estimated radiation dose amounts, the Fukushima Health Management Survey predicts there will be "no discernible increased incidence of radiation-related health effects" in the long term. Even if the eventual death toll from Fukushima's radiation reaches 500, as some studies predict, that would be a tiny number compared to the almost 20,000 lives taken by the tsunami itself.

Ironically, while the radiation itself seems to have done little damage to Fukushima's inhabitants, the threat of that radiation has caused incalculable harm to Tōhoku's reputation, and with it, Tōhoku's economy. Land values throughout the region plummeted in the wake of the disaster, wiping away billions of dollars in assets virtually overnight. While some manufacturing takes place in Tōhoku, the region is highly dependent on agricultural exports, especially rice, cattle, and dairy products, and demand for these products collapsed due to fears of radioactive contamination. Tōhoku's peach harvest, for instance, sold for only 30% of the prices it had commanded the year before. The same was true for Tōhoku's fishing industry, which has contracted enormously since 2010 not only due to destruction of the region's fishing fleet, but also due to lack of consumer demand for the region's seafood products, despite the fact that even fish caught in the waters adjacent to Fukushima Daiichi itself have contained no detectable levels of radiation since the mid-2010s. The slump in Tōhoku's fishing industry is likely to continue well into the future; indeed, as this chapter was being written, the planned release of water containing trace radioactivity levels from the Fukushima reactor site prompted China to declare a total ban of all Japanese seafood products. As for tourism, a third important industry in scenic Tōhoku, it also plummeted in the immediate aftermath of the disaster, though it began to rebound later in the decade as Tōhoku gained a reputation for "dark tourism." Visitors come to Tōhoku today to see disaster-related attractions such as the Great East Japan Earthquake and Nuclear Disaster Memorial Museum in Futaba and the Ukedo Elementary School in Namie, which was gutted by the tsunami and is now being preserved in that state as a memorial to the disaster.

The Fukushima disaster also struck a near-fatal blow to the Japanese nuclear power industry, though there are now signs of recovery. In the aftermath of the Fukushima Daiichi meltdown, a poll found that 80% of Japanese had turned against nuclear power, and bowing in part to public pressure, the government shut down all of Japan's reactors to review their safety. However, this decision came at a cost: energy prices for Japanese consumers rose nearly 20% from 2010 to 2013, and much more of Japan's electricity had to be generated by burning fossil fuels, undermining Japan's ability to reach greenhouse emissions reduction targets. Since 2013, public attitudes on nuclear power have shifted back toward nuclear power, due to higher energy prices combined with the lack of post-Fukushima radiation deaths, and as of the Fall of 2023 11 of Japan's nuclear plants had resumed operation. While Japan no longer plans to provide 50% of its electricity from nuclear power by 2030, nuclear power will play a sizable role in Japan's energy generation for the foreseeable future.

As of 2023, Tōhoku is just beginning to return to some degree of post-disaster normalcy, insofar as such a thing is possible after the loss of so many lives. Most of the 400,000 buildings destroyed have now been rebuilt or relocated, radiation levels are now negligible in most places, and much of the farmland contaminated by salt water has now recovered its fertility. What is more, to prevent a disaster on this scale from recurring in the future, coastal towns throughout Tōhoku have redoubled their tsunami protection systems. In Rikuzentakata, for example, the inadequate old seawall has been replaced by a more elaborate 12.5-meter structure, and the town site itself was raised up by 10 meters using rock fill, carried to the site from nearby hills using an elaborate system of conveyor belts. In total, Japan may have spent as much as $300 billion to rebuild the Tōhoku region – nearly four times Sri Lanka's annual GDP – which is a testimony to the resources that an industrialized nation like Japan can invest in post-disaster recovery and to reduce their vulnerability to natural hazards. However, whether Japan's improved defenses will withstand future tsunamis, especially in an age of global warming-elevated sea levels, remains to be seen.

A Final Note: Tsunamis and Xenophobia

It is worth pointing out that the Indian Ocean and Tōhoku Tsunamis were not followed by post-quake xenophobic violence, in sharp contrast to several of the earthquake case studies presented in Chapter 2. Far from further dividing the Indonesian government and the Aceh rebels, the Asian Tsunami actually created a moment for reconciliation between the two sides, reducing rather than exacerbating hostility. As for Sri Lanka, it is possible that the disease cues following the Asian Tsunami helped fuel the victory of anti-Tamil Sinhalese nationalists in the post-quake election, but that's just speculation.

In the case of the Tōhoku Tsunami, the evidence is mixed. While there was no genocide or widespread violence against scapegoated minorities, Japan did experience a spike of in-group preference and xenophobia following the 3/11 disaster. According to Tamaki Mihic (2020), right-wing hate speech and the use of derogatory, exclusionary language increased markedly in the first few years after 3/11. Much of this vitriol was directed against resident Chinese and Koreans living in Japan, which Mihic explicitly connects to similar xenophobic outbreaks which occurred after the 1995 earthquake in Hanshin and the 1923 Great Kantō earthquake discussed in Chapter 2. In addition, many Fukushima nuclear nomads faced discrimination and exclusionary behavior in the aftermath of the disaster. According to Tukio Yotsumoto and Shinichi Takekawa (2016), popular fears that Fukushima radioactivity could be transmitted from person to person led to instances of Fukushima evacuees being denied service at stores and medical facilities as well as alleged discrimination by employers, and these reported examples of discrimination are probably just the tip of the iceberg. Yotsumoto and Takekawa note that the marginalization of people rendered "impure" by radioactivity exposure is nothing new in Japan – the victims of Hiroshima and Nagasaki, for example, have also faced social exclusion and discrimination because of their perceived impurity.

In the end, the most likely explanation for the greater xenophobic reaction in Japan after 3/11, in comparison with the earlier Indian Ocean Tsunami, was the fact that during 3/11 the destruction of the tsunami waves was compounded by the release of radioactive fallout. Since the purpose of the behavioral immune system is pathogen avoidance, the BIS is sensitive to impurities and contamination, which trigger the emotion of disgust and elicit out-group exclusionary behavior. What is more, as discussed above, fears of radioactive contamination were already salient in Japanese culture due to the legacy of the American nuclear strikes in Hiroshima and Nagasaki, and this influenced how the Fukushima disaster was perceived by many Japanese. Thus, the greater xenophobic outbreak during the Tōhoku disaster compared to its Indian Ocean counterpart further illustrates one of the core themes of this book: the importance of understanding disasters within specific local contexts.

Further Reading

Barber, C. V. and K. Talbott. "The Chainsaw and the Gun: The Role of the Military in Deforesting Indonesia." *Journal of Sustainable Forestry*, vol. 16, no. 3 (2003), pp. 137–166.

Blackhall, Susan. *Tsunami*. London: Taj Books, 2005.

Fackler, Martin. "Crisis Saddles Village with Unwanted Notoriety." *New York Times*, April 5, 2011.

Gupta, Harsh K. and Vineet K. Gahalaut. *Three Great Tsunamis: Lisbon (1755), Sumatra-Andaman (2004), and Japan (2011)*. New York: Springer, 2013.

Hatsuzawa, Toshio and Takehiko Takano. "Characteristics of the Evacuation Area and the Spatial Distribution of the Radioactive Pollution in Fukushima Prefecture." In Pradyumna P. Karan and Unryu Suganuma (Eds.), *Japan after 3/11*. Lexington, KY: University of Kentucky Press, 2016, pp. 229–250.

Hyndman, Jennifer. "Siting Conflict and Peace in Post-Tsunami Sri Lanka and Aceh, Indonesia." In Piers M. Blaikie and Ragnhild Lund (Eds.), *The Tsunami of 2004 in Sri Lanka*. New York: Routledge, 2010, pp. 88–95.

Karan, Pradyumna P. "Tami Nadu and Tohoku: The Two Tsunamis." In Pradyumna P. Karan and Unryu Suganuma (Eds.), *Japan after 3/11*. Lexington, KY: University of Kentucky Press, 2016, pp. 447–461.

Karan, Pradyumna P. and Unryu Suganuma (Eds.) *Japan after 3/11*. Lexington, KY: University of Kentucky Press, 2016.

Krauss, Erich. *Wave of Destruction: The Stories of Four Families and History's Deadliest Tsunami*. Emmaus, PA: Rodale, 2006.

Kumagai, Atsushi and Koichi Tanigawa. "Current Status of the Fukushima Health Management Survey." *Radiation Protection Dosimetry*, vol. 182, no. 1 (2018), pp. 31–39.

Lochbaum, David, Edwin Lyman, and Susan Q. Stranahan. *Fukushima: The Story of a Nuclear Disaster*. New York: The New Press, 2014.

Lutgens, Frederick K. and Edward J. Tarbuck. *Foundations of Earth Science*, 7th ed. New York: Pearson, 2014.

Molesky, Mark. *This Gulf of Fire: The Destruction of Lisbon*. New York: Alfred A. Knopf, 2015.

Nadesan, Majia Holmer. *Fukushima and the Privatization of Risk*. New York: Palgrave Macmillan, 2013.

Nazara, S. and B. P. Resosudarmo. "Indonesia: The First Two Years after the Tsunami." In S. Jayasuriya and P. McCawley (Eds.), *The Indian Ocean Tsunami: Aid and Reconstruction after a Disaster*. Northampton, MA: Edward Elgar Publishing, 2010, pp. 68–122.

Papadopoulos, Gerassimos A. "Two Large Tsunamis in the Prehistory of the Aegean Sea: The Minoan Tsunami (~17th Century BC) and the Troy Tsunami (~13th Century BC)." In G. A. Papadopoulos and K. Stake (Eds.), *Proceedings of the 22nd International IUGG Tsunami Symposia* (2005), pp. 181–185.

Papadopoulos, Gerassimos A. *et al.* "Historical and Pre-Historical Tsunamis in the Mediterranean and its Connected Seas." *Marine Geology*, vol. 345 (2014), pp. 81–109.

Parwanto, Novia Budi and Tatsuo Oyama. "Investigating the Impact of the 2011 Great East Japan Earthquake and Evaluating the Restoration and Reconstruction Performance." *Journal of Asian Public Policy*, vol. 8, no. 3 (2015), pp. 329–350.

Prager, Ellen. *Furious Earth: The Science and Nature of Earthquakes, Volcanoes, and Tsunamis*. New York: McGraw-Hill, 2000.

Reilly, Benjamin. *Disaster and Human History: Case Studies in Nature, Society, and Catastrophe*. 2nd ed. Boone, NC: McFarland, 2022.

Sathiparan, Navaratnarajah. "An Assessment of Building Vulnerability to a Tsunami in the Galle Coastal Area, Sri Lanka." *Journal of Building Engineering*, vol. 27 (2020). doi:10.1016/j.jobe.2019.100952.

Stradford, H. Todd. "Earthquake and Tsunami in Taro Town." In Pradyumna P. Karan and Unryu Suganuma (Eds.), *Japan after 3/11*. Lexington, KY: University of Kentucky Press, 2016, pp. 138–159.

Suganuma, Unryu. "Historical Geography of the Japanese Tsunami." In Pradyumna P. Karan and Unryu Suganuma (Eds.), *Japan after 3/11*. Lexington, KY: University of Kentucky Press, 2016, pp. 45–73.

Suganuma, Unryu. "TEPCO and Nuclear Energy Politics: An Analysis of the 'Japanese Pentagon'." In Pradyumna P. Karan and Unryu Suganuma (Eds.), *Japan after 3/11*. Lexington, KY: University of Kentucky Press, 2016, pp. 204–228.

Sugawara, Daisuke *et al.* "The 2011 Tohoku-oki Earthquake Tsunami: Similarities and Differences to the 869 Jogan Tsunami on the Sendai Plain." *Pure and Applied Geophysics*, vol. 170 (2012), pp. 831–843.

Taylor, Jonathan. "Ramifications of the Fukushima Nuclear Disaster: Toward the End of the 'Peaceful Atom'?" In Pradyumna P. Karan and Unryu Suganuma (Eds.), *Japan after 3/11*. Lexington, KY: University of Kentucky Press, 2016, pp. 189–203.

Yamazaki, Kenji and Tomoko Yamazaki. "Disaster Prevention Culture: Role of Schools in Saving Tsunami Victims." In Pradyumna P. Karan and Unryu Suganuma (Eds.), *Japan after 3/11*. Lexington, KY: University of Kentucky Press, 2016, pp. 331–363.

Works Cited

Birmingham, Lucy and David McNeill. *Strong in the Rain: Surviving Japan's Earthquake, Tsunami, and Fukushima Nuclear Disaster.* New York: Palgrave Macmillan, 2012.

Blaikie, Piers M. and Ragnhild Lund (Eds.) *The Tsunami of 2004 in Sri Lanka: Impacts and Policy in the Shadow of Civil War.* New York: Routledge, 2010.

Esteban, Miguel *et al.* "Analysis of Tsunami Culture in Countries Affected by Recent Tsunamis." *Procedia Environmental Sciences*, vol. 17 (2013), pp. 693–702.

Gamburd, Michele Ruth. *The Golden Wave: Culture and Politics after Sri Lanka's Tsunami Disaster.* Bloomington, IN: Indiana University Press, 2014.

Jasparro, Christopher and Johnathan Taylor. "Transnational Geopolitical Competition and National Disasters: Lessons from the Indian Ocean Tsunami." In Pradyumna P. Karan and Shanmugam P. Subbiah (Eds.), *The Indian Ocean Tsunami.* Lexington, KY: University Press of Kentucky, 2011, pp. 283–299.

Karan, Pradyumna P. and Shanmugam P. Subbiah (Eds.) *The Indian Ocean Tsunami: The Global Response to a Natural Disaster.* Lexington, KY: University Press of Kentucky, 2011.

Khasalamwa, Sarah. "Is 'Build Back Better' a Response to Vulnerability? Analysis of the Post-Disaster Humanitarian Interventions in Sri Lanka." In Piers M. Blaikie and Ragnhild Lund (Eds.), *The Tsunami of 2004 in Sri Lanka.* New York: Routledge, 2010, pp. 72–86.

Klein, Naomi. *The Shock Doctrine: The Rise of Disaster Capitalism.* London: Picador, 2008.

Mahaffey, James. *Atomic Accidents: A Story of Nuclear Meltdowns and Disasters.* New York: Pegasus Books, 2014.

Mihic, Tamaki. *Re-Imagining Japan after Fukushima.* Canberra, Australia: Australian National University Press, 2020.

Minoura, K. *et al.* "The 869 Jōgan Tsunami Deposit and Recurrence Interval of Large-Scale Tsunami on the Pacific Coast of Northeast Japan." *Journal of Natural Disaster Science*, vol. 23, no. 2 (2001), pp. 83–88.

Mulligan, Martin and Judith Shaw. "Achievements and Weaknesses in Post-Tsunami Reconstruction in Sri Lanka." In Pradyumna P. Karan and Shanmugam P. Subbiah (Eds.), *The Indian Ocean Tsunami.* Lexington, KY: University Press of Kentucky, 2011, pp. 237–260.

Nakamura, Fuyubi. "Memory in the Debris: The 3/11 Great East Japan Earthquake and Tsunami." *Anthropology Today*, vol. 28, no. 3 (2012), pp. 20–23.

Nicol, Bill. *The Tsunami Chronicles: Adventures in Disaster Management.* Bill Nicol, 2013.

Parry, Richard Lloyd. *Ghosts of the Tsunami: Death and Life in Japan's Disaster Zone.* New York:MCD/Farrar, Straus and Giroux, 2017.

Rambelli, Fabio. "Gods, Dragons, Catfish, and Godzilla: Fragments for a History of Religious Views on Natural Disasters in Japan." In Roy Stars (Ed.), *When the Tsunami Came to Shore: Culture and Disaster in Japan.* Leiden: Brill, 2014, pp. 50–69.

Saatcioglu, M. A., Ghobarah, and I. Nistor. "Performance of Structures Affected by the 2004 Sumatra Tsunami in Thailand and Indonesia." In Tad S. Murty, U. Aswathanarayana, and N. Nirupama (Eds.), *The Indian Ocean Tsunami.* New York: Routledge, 2007, pp. 297–321.

Samuels, Annemarie. *After the Tsunami: Disaster Narratives and Remaking of Everyday Life in Aceh*. Honolulu, HI: University of Hawaii Press, 2019.

Silva, Kalinga Tudor. "'Tsunami Third Wave' and the Politics of Disaster Management in Sri Lanka." In Piers M. Blaikie and Ragnhild Lund (Eds.), *The Tsunami of 2004 in Sri Lanka*. New York: Routledge, 2010, pp. 60–71.

Umitsu, Masatomo. "The Geoenvironment and the Giant Tsunami Disaster in the Northern Part of Sumatra Island, Indonesia." In Pradyumna P. Karan and Shanmugam P. Subbiah (Eds.), *The Indian Ocean Tsunami*. Lexington, KY: University Press of Kentucky, 2011, pp. 51–63.

Umitsu, Masatomo. "Tsunami Flow and Geo-Environment of the Pacific Coastal Region of Tohoku." In Pradyumna P. Karan and Unryu Suganuma (Eds.), *Japan after 3/11*. Lexington, KY: University of Kentucky Press, 2016, pp. 104–137.

Van Dam, Arvid. "Negotiating the Indian Ocean: Opportunities in the Process of Recovery in Post-Tsunami Sri Lanka." *Etnofoor*, vol. 27, no. 1 (2015), pp. 37–52.

Yamazaki, Kenji and Tomoko Yamazaki. "Tsunami Disasters in Seenigama Village, Sri Lanka, and Taro Town, Japan." In Pradyumna P. Karan and Shanmugam P. Subbiah (Eds.), *The Indian Ocean Tsunami*. Lexington, KY: University Press of Kentucky, 2011, pp. 135–159.

Yotsumoto, Yukio and Shinichi Takekawa. "The Social Structures of Victimization of Fukushima Residents Due to Radioactive Contamination from the 2011 Nuclear Disaster." In Pradyumna P. Karan and Unryu Suganuma (Eds.), *Japan after 3/11*. Lexington, KY: University of Kentucky Press, 2016, pp. 251–268.

4 Volcanoes

On July 20, 1783, Reverend Jón Steingrísson gathered his small flock of Icelandic farmers and herders into the parish church of Klaustur for what many feared might be their last Sunday mass. In the preceding weeks, the ground had trembled with earthquakes, the air had filled with ash and blue smoke, and plumes of fire had spouted in the distance. Worst of all, a river of lava was snaking its way down the empty bed of the Skaftá gorge, flowing inexorably toward the town. Two upstream churches had already been consumed by lava, which was advancing up to 6 kilometers a day, and Klaustur seemed to be next.

In the face of the advancing lava, Steingrísson marshaled the town's spiritual defenses. As lightning generated by roiling ash clouds flashed overhead, Steingrísson beseeched his assembled parishioners to ask God, "with proper meekness," to spare Klaustur and its church. And, seemingly, their prayers were answered. When the inhabitants of Klaustur checked the progress of the lava after this "Fire Mass," as it came to be called, they found that the river of lava had stopped, miraculously, just a few kilometers from Klaustur.

As it turned out, however, Klaustur's troubles were far from over. The earthquakes ended and the distant fires dwindled as summer gave way to winter, but a pall of blue smoke remained, and continued ashfalls destroyed the crops in the fields and killed or drove away the fish in nearby lakes and streams. Most concerningly of all, the livestock began to suffer ghastly symptoms of illness. "Hard, swollen lumps grew from their joints," Alexandra Witze and Jeff Kanipe (2014) report, and "their heads became swollen and their jaws so weak they could barely graze … tails and hooves fell off, ribs became warped, and the animals' hair dropped out in patches." As Steingrísson noted in his diaries, the meat of these animals proved to be "foul-smelling and bitter and full of poison, so that many a person died as a result of eating it."

Steingrísson's parishioners ate this tainted meat because they had no choice – they were starving. During Iceland's *Móduhardindin* famine, which means "hardship of the fog," hungry Icelandic villagers resorted to

DOI: 10.4324/9781003436805-5

chewing on leather or consuming pseudo-foods, such as porridges made of fine-chopped hay or ground-up fish bones. Worst still, the same symptoms that afflicted the livestock began to appear in Steingrísson's human flock, including the growth of "ridges, growths, and bristle" on their ribs, joints, legs, and hands. "Their bodies became bloated," Steingrísson recorded in his diaries, "the insides of their mouths and their gums swelled and cracked, causing excruciating pains." Some even lost their teeth and their tongues to this malady, which we now know was probably fluorine poisoning. In the end, Steingrísson had to preside over the burials of 76 of his parishioners over the course of 1784.

Iceland was far from alone in its misery. As we will see below, nearly the entire world suffered during this *annus mirabilis*, or "year of wonders," which was triggered by the release of enormous quantities of sulfates and other gases during the Laki volcano eruption. Excessive heat and bitter cold ruined harvests from Europe to Japan, monsoon failures brought hunger to India, and meager Nile floods triggered a famine that may have killed or driven away as many as a sixth of Egypt's inhabitants. The Laki eruption, therefore, epitomizes the double threat of volcanic hazards, which can kill both directly through eruptive impacts and indirectly through rapid short-term climate change.

The Science of Volcanoes

Volcano Origins

Volcanoes, like earthquakes and most tsunamis, result from the processes of plate tectonics, the slow movement of the Earth's crustal plates as they are dragged along by convection currents in the semi-fluid mantle. As a result, like earthquakes and tsunamis, volcanoes are highly localized phenomena, much more common in some regions (and along some types of plate boundaries) than others. Approximately 80% of all volcanoes, for example, lie underwater along the mid-ocean ridges, which are constantly building new seafloor in the Pacific, Atlantic, and Indian Oceans. These volcanoes form at "normal" faults, where two plates are moving apart from each other, creating an avenue for upwelling magma. In contrast, volcanoes never form at "transform" faults, where two plates are moving side-by-side, since plate boundaries of this sort create no route for magma to reach the surface.

As for "thrust" faults, where two plates are colliding directly, their relationship to volcanism is more complex. As we already discussed in Chapter 1, if the two colliding plates are both continental plates, they will buckle and deform into mountain ranges, such as the Alps and the Himalayas. Volcanoes are rare along such plate boundaries as they offer no avenue for magma to reach the surface. If one of the plates is an oceanic plate, however, something interesting happens: water and gases carried into the upper mantle by the descending plate change the chemistry of the magma below the lip of the continental plate, lowering its melting point and making it more buoyant. As a result, magma bubbles upward through the continental plate, carrying dissolved water and carbon dioxide with it, and generally also picking up some silica along the way as it partially melts the silica-rich rocks of the continental plate. Eventually, this rising magma puddles together in a magma chamber under the continental plate, and because it is lighter than the surrounding rock, it seeks a route to the surface.

If this gas and silica-rich magma does find an escape route, the resulting volcano is typically a stratovolcano. Such volcanoes are also known as "composite" volcanoes, since their slopes generally consist of alternating layers of hardened lava and pyroclastic materials, a general term for any rock fragment ejected into the air as a volcano erupts. Pyroclasts, which means "broken by fire," can range from tiny ash particles, which are essentially miniscule pieces of vaporized rock, to large "lava bombs" the size of automobiles. Stratovolcanoes can also be distinguished from other volcanoes, especially the relatively flat "shield" volcanoes that form at normal faults, by their steep slopes, as epitomized by the graceful contours of the well-known Mount Fuji (Figure 4.1). The reason for the more pronounced slopes of stratovolcanoes is the higher viscosity of their magma due to its high silica content, so it tends to cool and harden before traveling very far from the volcanic vent from which it emerged.

Stratovolcanoes are different from the volcanoes at normal faults in another way as well: because their magma contains a good deal of dissolved water and gases, stratovolcanoes are prone to violent, explosive eruptions. Jelle de Boer and Donald Sanders (2002) liken this process to the chemistry of a "popping" bottle of champagne. Like the magma of a stratovolcano, champagne contains high levels of dissolved gases, principally carbon dioxide, but so long as the bottle remains sealed, pressure keeps that gas within the liquid solution. Remove the cork, however, and the gas separates from the fluid and explodes out of the bottle, carrying some of the champagne with it in a celebratory froth. Similarly, when a stratovolcano pops its top, gases suspended in the magma explode outwards, in the process "shredding the molten magma into myriad droplets that, upon cooling, become pyroclastic fragments." A volcano is especially prone to violent eruptions if the magma has high viscosity and thus offers some resistance to the release of the gases, and as mentioned above, stratovolcanoes tend to feature thick, viscous magma because of their high silica content.

A final mechanism that generates volcanoes is mantle plumes – strong magma convection cells under the Earth's crust which create "hot spots" capable of melting through the tectonic plate above. The best-known example of hot spot volcanism is the Hawaiian Islands, where the Pacific Plate has been passing over a mid-plate hot spot for millions of years, creating a chain of islands formed by magma welling up from below, a process that could be likened to the line of drip marks left behind on the

Figure 4.1 View of Mount Fuji from Koshigaya, Japan
Source: One of the 36 views of Mount Fuji by Katsushika Hokusai, courtesy of the Metropolitan Museum of Art.

asphalt by a moving car with an oil leak. Thankfully for those who live in the shadow of hot spot volcanism, the magma here tends to be low in carbon dioxide, water, and silica. As a result, when this magma arrives as lava at the Earth's surface, it flows freely, as was the case in the 1783–1784 Laki eruption in Iceland that started this chapter, but usually not explosively.

Volcanic Hazards

In simple terms, volcanoes can kill you in two distinct ways: through physics, and through chemistry. Those in the vicinity of a volcano may fall victim to cascading volcanic debris, superheated pyroclastic flows, and post-quake lahars (which are discussed below) and occasionally localized tsunami waves. Volcanoes can also kill chemically, either directly through asphyxiation or fluorine poisoning, or indirectly by changing the chemistry of the atmosphere, triggering short-term cooling and other climate change.

One hazard of the volcano is the explosion itself, which results when pressurized magma full of dissolved gases finds a way into the atmosphere, for example, by exploiting a weakness created by a landslide on the mountain face or by dislodging a lava plug. Like atomic bombs, volcanic explosions are often measured in megatons of TNT, and even mid-range volcanic eruptions like that of Mt. Saint Helens in 1980 can release more than 25 kilotons of energy, the equivalent of over 1,500 Hiroshima nuclear explosions. If these volcanic explosions strike a body of water, they can trigger a tsunami, as we will see below in the case of the Krakatau eruption.

While volcanic explosions can be immensely destructive over short distances, a greater physical threat posed by volcanoes is pyroclastic flows, destructive clouds of superheated, vaporized magma that can travel faster than the speed of sound. In the case of Mt. Saint Helens, the pyroclastic flow spread out over a fan-shaped area that extended up to 31 kilometers away from the volcano. Trees struck by pyroclastic flows are either vaporized, knocked flat, or scorched to death, depending on the distance from the volcano, while buildings impacted by pyroclastic flows are simultaneously toppled and set on fire. As for humans, those unfortunate enough to be caught in a pyroclastic flow either die instantly from the heat or impact, or succumb to suffocation after the inhalation of superheated rock burns their lungs from the inside. Most of the 57 people known to have died during the Mt. Saint Helens disaster were killed by pyroclastic flows.

The pyroclastic materials left behind by these eruptions, in turn, contributes to a third volcanic hazard: lahars, or mudslides, which result when recently-deposited pyroclastic materials that have been loosened by water slump downhill, carrying away everything in their path. Although the term "lahar" is Indonesian, lahars are a truly global hazard. During the Mt. Saint Helens disaster, for example, the almost instantaneous melting of the mountain peak's glaciers and snow triggered a mudslide that traveled as fast as 140 kilometers an hour and traveled 50 of more kilometers downstream along river and creek beds. What is more, in the years since the eruption, new lahars have occurred almost every time Mt. Saint Helens' debris is soaked with heavy rains or spring meltwater. As recently as May 2023, a lahar near Mt. Saint Helens rushed down South Coldwater Creek and destroyed a bridge on a state highway. No one was killed, but 11 people were stranded and had to be rescued by helicopter. Once again, keep in mind that the Mount St. Helen's eruption was only moderate by global standards. The potential for death and destruction due to the direct eruptive impact, the pyroclastic flows, and the lahars is exponentially greater during the larger eruptions, such as the events considered in the case studies below.

In addition to these straightforward physical hazards, volcanoes can kill through chemical processes as well. As mentioned above, volcanoes can emit large amounts of fluorine, an element that illustrates the axiom that "the dose makes the poison." While low levels of fluorine are good for dental health, high levels of fluorine can lead to bone deformations, including the symptoms experienced in Klaustur during the Laki eruption, in addition to death. Volcanic activity has also been implicated in "limnic eruptions," a rare but potentially fatal phenomenon where carbon dioxide from volcanic gases or other sources builds up in the lower waters of a lake. Because their waters are essentially carbonated, these lakes can be highly unstable and can raise dramatically in height if the CO_2 is released, spilling out like the contents of a shaken soda can. In addition, the sudden release of CO_2 into the surrounding area displaces normal, breathable air, sometimes leading to mass asphyxiation events. During the well-studied limnic eruption in Lake Nyos in Cameroon, for example, the lake shore was swept by a 25-meter wave; worse yet, the sudden release of hundreds of thousands of tons of carbon dioxide killed over 1,700 people living nearby, many of whom died in their sleep after inhaling this invisible but deadly gas.

Volcanoes can also impact historical processes by changing the weather, mainly by altering the chemical composition of the Earth's atmosphere. Carbon dioxide, a greenhouse gas which is released in large amounts during volcanic eruptions, contributes to global warming, and volcanic CO_2 emissions are now thought to have helped end the last "snowball Earth" phase of world geological history 635 million years ago, thawing the planet and thereby making it more favorable to life. Most modern global warming, however, is the result of human rather than volcanic activities. As U.S. Geological Survey scientist Terry Gerlach pointed out in a 2011 article, Mt. Saint Helens released 10 megatons of carbon dioxide into the atmosphere during its eruption, but today it takes human society just 2.5 hours to release the same amount.

The most important volcanic gas in terms of climate change, however, is sulfur dioxide. When released into the atmosphere, SO_2 mixes with water vapor to form sulfuric acid aerosols, which are highly reflective to solar radiation, leading to significant short-term cooling of the planet's surface. While Mt. Saint Helens produced relatively little sulfur, the Mount Pinatubo eruption 11 years later released enough sulfates to cause global temperatures to drop by about .5° C, and this effect lasted for the next three years. As we will in the final section of this chapter, even stronger historical eruptions have led to more pronounced episodes of global cooling, such as that which occurred during the Mount Toba eruption of 74,000 years ago, the unknown eruption or eruptions that produced the climate event of 536 CE, the Laki eruption of 1783–1784, and the Tambora eruption of 1816, which triggered the "year without a summer" and the last great subsistence crisis of the western world.

Because of the dual nature of their hazards, volcanoes can be said to be both among the least deadly and the most deadly of natural disasters. In terms of direct death toll, volcanoes only kill a fraction as many people each year as earthquakes, floods, and tsunamis. At the same time, the death toll caused by volcano-induced climate changes can reach into the hundreds of thousands or millions, to the point that some experts believe that the Laki eruption might have been the deadliest natural disaster in known human history.

Volcanoes and Culture

Given their terrifying hazards, it is little surprise that volcanoes have left their mark upon the culture of societies living in volcanic zones. Many religious systems associate volcanism with the gods, including the ancient Etruscans who lived in volcanically-active South Italy and who worshiped Velkhan as the god of destruction, fire, and the hearth. Velkhan would be later be adopted by the Romans as Vulcan and assigned the job of metal-worker for the gods – the rumblings of the mountains, the Romans thought, was a sign that Vulcan was swinging his blacksmith's hammer. Japan's eight active volcanoes, in turn, were associated in traditional Shinto religion with the goddess Konohanasakuya-hime, whose name means princess of the blooming cherry-tree blossom. Interestingly, she is regarded not as a goddess of destruction but of life and growth, which might reflect the important role Japan's volcanoes have played in both building this archipelago of islands and sustaining the fertility of its soil. Similarly, R. J. Blong's (1982) work on agriculturalist societies in New Guinea notes that some tribal groups have developed elaborate religious ceremonies to try to encourage the fall of volcanic ash, knowing it will increase their crop yields.

Other cultures took a dimmer view of volcanism, and associated the eruption of volcanoes with divine wrath, anger which could only be turned aside through supernatural protection. Medieval Christians, for example, believed that volcanoes were gateways to hell and appealed to saints, such as the martyred Saint Agatha, for her intercession on their behalf. Similarly, archeologists in Mexico believe that the numerous effigies and shrines found in Aztec villages on the slopes of volcanic Mount Popocatépetl were an attempt to appease the mountain and prevent its eruption. In this case, divine protection proved to be ineffective: the Aztec shrines and villages were discovered beneath a preserving layer of volcanic pyroclasts. Jeanet Bentzen (2019) has argued that the presence of volcanoes within 1,000 kilometers correlates significantly with a society's overall levels of religious belief and observance, though to a lesser degree than earthquakes and tsunamis. For Bentzen, the trait that links volcanism to religiosity is the unpredictability of volcanic eruptions: human beings are more likely to seek divine assistance in the case of occasional and unforeseeable disasters.

In the West, the scientific study of volcanism was long hampered by the Aristotelian idea that the planet Earth, like the element earth, was cold and dry, ruling out the possibility of a molten liquid interior. Volcanic eruptions, therefore, must be the result of accidental factors. As Rienk Vermij (1998) points out, Aristotelians believed volcanism was the result of the ignition of "large masses of sulfur in the earth's interior," which would eventually burn itself out once the fuel was consumed. Once gunpowder was adopted by the West, in fact, some European scholars expanded upon this idea and proposed that volcanoes and quakes were the result of underground explosions caused by the admixture of naturally-occurring charcoal, saltpeter, and sulfur! Ironically, the Christian belief that volcanic fire was hellfire, and that hell was located within the Earth's interior, was actually closer to the truth than Aristotle's "scientific" views. It should be noted that this linkage between volcanoes, hellfire, and divine punishment has by no means vanished even today. As David K. Chester and Agnus M. Duncan (2009) have pointed out, nearly every volcanic eruption of the twentieth century was interpreted as the result of divine retribution, including the 1980 eruption of Mt. Saint Helens, which some radio evangelists portrayed as "a warning from god about the evils of drinking strong liquor."

Turning finally to "cultures of volcanism" – cultural adaptation to volcanic hazards – the record of such adaptations is fairly sparse, since the sheer scale of the hazards that volcanoes pose makes preparing against them almost impossible. Volcanic eruptions after all pack a punch best measured in megatons of TNT, the same assessment scale used for atomic weaponry. The most obvious means to avoid such hazards is to live far away from the slopes of a volcano. In practice, however, volcanic zones are often thickly settled, thanks to their rich soils, fertilized by volcanic ash. What is more, because stratovolcanoes are the result of the subduction of oceanic plates under continental plates, they tend to form close to the sea, and coastlines are often densely settled because of their proximity to marine resources and long-distance trade.

Nonetheless, human beings can take some precautions to reduce volcanic hazards. According to a study by R. Funiciello et al. (2003), the area near Rome was subjected to repeated floods by the Albano Lake in the past, which were likely the result of limnic eruptions, since the Albano Hills are active volcanoes that are constantly venting CO_2 gas. While the ancient Romans ascribed these lake eruptions to the anger of the gods, they took practical steps to prevent them from recurring, building a drain-and-tunnel system through the side of the hill to lower the level of the lake. As K. V. Cashman and G. Giordano (2008) point out, this lake drainage project may represent the world's first "engineering-based volcanic hazard mitigation measure," and they note it predates by 2,000 years the construction of a similar structure built by the Dutch to control the water levels in the Kelut crater lake in Indonesia in 1912. In recent years, engineers have turned their attentions to another volcanic hazard, lahars, which can be brought under control through the construction of sediment capture structures. Case in point is the retention dam on the North Fork Toutle River on the slopes of Mt. Saint Helens, designed to prevent destructive lahars from penetrating further down the river. As we will see below, a similar dam project was put in place in the Philippines after the Pinatubo eruption, but with limited success.

The most widespread mitigation measure used today against volcanic hazards, however, is early detection and evacuation. Modern vulcanology has developed several techniques to predict upcoming eruptions, including seismographs, thermal imaging cameras, and tiltmeters, sensors used to detect growing deformations in the mountain slopes which indicate pressure is building up within the volcano. As a result, the death toll from volcanic hazards in the modern era has steadily declined relative to less predictable hazards such as earthquakes and tsunamis.

Case Studies in Immediate Impact Volcanism

Since volcanic hazards can impact human populations in two distinct ways – directly, through physical effects, and indirectly by triggering rapid short-term climate change – the rest of the chapter will be divided into two broad sections that consider each type of volcanic impact in turn. Keep in mind, however, that these two categories are by no means mutually exclusive. Some eruptions, like the 1991 Mount Pinatubo disaster, had both dramatic close-range impacts and wider-ranging impacts on global climate, since the sulfates exhaled by Pinatubo reduced the world's average temperature by about .5°C in the years after the eruption. However, since the main goal of the Pinatubo section below is to examine the vulnerability and resilience of the Aeta, a vulnerable and marginalized Indigenous population in the Philippines, only Pinatubo's local effects will be discussed below.

Krakatau, 1883

The Krakatau eruption has two traits that lend themselves well to the study of volcanic hazards. Not only was it the second-strongest eruption of the nineteenth century – only Tambora, described below, was stronger – it was also extremely well documented, as Krakatau lay along a busy global shipping lane and was only 160 kilometers from Batavia, the capital of the Dutch East Indies. A stratovolcano created by the subduction of the Indo-Australian plate beneath the Eurasian plate, Krakatau's tapered cone was well known to sailors, who frequently passed this isolated volcanic peak when traversing the Sunda Straits that separate Sumatra from Java. Both Indonesian islands were then under the control of the Dutch East Indies, though Dutch control over Aceh in northern Sumatra was still uncertain, contested by an ongoing Acehnese guerilla campaign.

Although the island of Krakatau itself was uninhabited, the volcano overlooked a region that was both well populated and highly vulnerable to volcanic hazards. Both sides of the Sunda Straits were lined with a low-lying alluvial coastal plain that stretched up to 6–8 kilometers inland, and according to Alwin Scarth (1999), this densely inhabited region was covered by both an "intricate patchwork" of rice and cane fields as well as "hundreds of kampongs, or villages" consisting of "flimsy huts made of rattan, wickerwork, and thatched with palm leaves." The relatively few Dutch who lived in the region were concentrated in a few larger towns, including Anjer on Java and Telok Betung on the Sumatran coast. While the Java coast was relatively straight and unindented, the Sumatran coastline suffered from additional vulnerabilities in the form of two deep, enclosed bays that resembled the *ria* of Japan, and like the *ria*, would concentrate and magnify some of Krakatau's volcanic effects. In addition to occupying a precarious geographic location, the local Indonesian population was economically vulnerable as well, groaning under the weight of heavy Dutch taxation that amounted to a fifth of the crop and bound them to their villages by harsh labor control regulations. Nonetheless, there is reason to suspect that, in the case of this disaster at least, the economic vulnerability of the local population played little role in mediating the impact of the Krakatau disaster. Krakatau's hazards were of such a magnitude that they disregarded normal socioeconomic and racial boundaries, striking rich and poor, Indonesian and Dutch alike.

These hazards impacted a population that was almost completely unprepared for volcanic hazards. The Dutch in particular had little knowledge or understanding of volcanic events, regarding Krakatau's rumbles and smoke plumes as a curiosity rather than a threat, and they even organized an "agreeable excursion" to the volcano on May 27, three months before Krakatau exploded. While the Dutch tended to live at higher elevations along the coast than native Indonesians, this reflected a fear of the disease-causing "miasmas" of Indonesia's fishponds, mangroves, and mudflats rather than a precautionary measure against volcanic hazards. As for the Indonesian natives, they were more knowledgeable about volcanism, which is hardly surprising, since Indonesia boasts more volcanoes than anywhere else on Earth. Although most Indonesians were nominally Muslims at the time, the syncretic Islam then practiced in Indonesia still had room for pre-Islamic deities, including volcano gods like Krakatau's Orang Alijeh, who signaled his anger at mankind by "spewing forth fire and gas and lava." What is more, according to Simon Winchester (2003), many Indonesians still regarded volcanic eruptions as "astral messages," warnings which mortals ignored at their peril.

Nonetheless, Indonesia's culture of volcanism did not translate into effective anti-volcanic precautions and preparations. It may be that spiritual protection was the only form of protection available to the native Indonesians: given their lack of resources and the scale of the potential hazards, it is hard to imagine what practical steps they could have taken that might have lessened the impact of the 1883 disaster.

On August 27th, 1883, the hazards spawned by Krakatau's eruption combined with the vulnerability of the local populations to kill over 36,000 people on the shores of the Sunda Straits. On the morning of that day, the Krakatau volcano released a series of thunderous explosions that were heard as far as 4,775 kilometers away. These blasts triggered several tsunamis, including one strong enough to overtop 35-meter hills along the nearby coasts. While the tsunamis might have been caused by the explosion itself, some scholars believe they were the result of ocean displacement by Krakatau's powerful pyroclastic flows.

Whatever their cause, these waves inflicted almost indescribable damage on the coasts of Java and Sumatra. One Dutch observer described the incoming tsunami as resembling a "low range of hills rising out of the sea," and when it receded, it carried away "the dead bodies of many a friend and neighbor." These monstrous waves left behind a landscape fundamentally transformed; according to one sailor on a passing ship, "the shores on each side [of the Sunda Strait] looked burnt up and sterile. Java is known as the Garden of the East, but this bit looked more like the Sahara, absolutely desolate." Closer inspection revealed a mangled coastline, with palm trees torn up from the roots and toppled, "lying in endless confusion one above another." As for the native houses, "made of their frail materials of bamboo and leaves," their debris covered the ground, along with "all kinds of smashed furniture, broken cooking utensils, doors wrenched from their hinges, and every article of native costume in one great indescribable mass." The entire coastal plain, British observer Philip Neale proclaimed, was "so completely ruined as to be nothing more or less than a huge cemetery" (Simkin and Fiske, 1983). Some of the villages on the Javan coast were never rebuilt, despite the fertility of the local soil, for fear of future Krakatau eruptions.

While Krakatau's tsunamis accounted for the bulk of the casualties, a few unlucky souls were close enough to the volcano to feel the searing impact of Krakatau's pyroclastic flows. The Dutch-born Beyernick family, who at the time of the eruption were living only about 24 kilometers from Krakatau near the southern tip of Sumatra, had fled the coastal lowlands along with thousands of native Indonesians and were waiting out the eruption in a hut on Rajabasa Mountain. On the day of the eruption, Mrs. Beyernick remembered, the world outside the hut was suddenly thrown into pitch-black darkness, and volcanic ash began to rush up "like a fountain" through the floorboards of the hut. Mrs. Beyernick felt heavy pressure, "as if all the air was being sucked away," and winced as the hot ash particles "pricked like needles," scorching her skin. She eventually fled the hut into the darkness outside, doubled over with pain. As she walked, she felt as if thorns or hooks were pulling everywhere on her skin, and eventually came to the horrifying realization that much of her skin was hanging off her body in strips, since she had been badly burned by Krakatau's pyroclastic flow. Nor was she alone; almost all of the 3,000 Indonesians and Dutch who had fled to Rajabasa suffered burns and blisters, and 1,000 would die of their injuries. In the Dutch East Indies, as a whole, about 4,000 would be killed by the pyroclastic flow, mainly along the Sumatran coast.

As for Krakatau itself, not much was left; as a telegram between Batavia and Singapore put it the day after the explosion, "where once Mount Krakatau stood, the sea now plays." One sailor described the same scene with less delicate language: "the island," he wrote, "had spit itself out." Fully 70% of the mountain was gone, or more precisely, was scattered over the nearby seascape, reshaping the seafloor and creating navigational problems for years.

Thanks to the telegraph, news of the Krakatau disaster quickly reached the outside world, triggering an outpouring of humanitarian aid for the victims of the disaster. Nonetheless, the eruption seems to have heightened tension between the Dutch and the Indonesians. Many Indonesians blamed the Dutch for the disaster, seeing it as divine retribution for the colonial government's ongoing war against the Aceh Sultanate. Then, just five weeks after the disaster, a Dutch soldier was stabbed repeatedly in a marketplace in Java by a bearded Indonesian assailant wearing the white robes associated with the *Hajj*, the holy pilgrimage to Mecca. This attacker got away, but six weeks later, another knife-armed assailant in white robes was arrested while trying to infiltrate a Dutch garrison, though not before he had wounded an Indonesian soldier on guard duty at the garrison headquarters.

The Dutch would blame these attacks on religious "fanaticism" brought back from Mecca by returning *Hajjis*, and there is probably some truth to this, though frustration with the Dutch East Indies' exploitative economic policies no doubt played a role as well. Nonetheless, the fact these attacks occurred in the immediate aftermath of the eruption, when disease cues were still ubiquitous and many bodies were yet unburied, suggests that the behavioral immune system might have played a role in predisposing post-disaster Indonesians toward xenophobic behavior. If so, the psychological impact of the Krakatau eruption may have contributed to the steady rise of anti-Dutch sentiment during the second half of the 1880s, which culminated in the failed Banten Peasant Revolt of 1888 against the Dutch occupation of Java.

Mount Pinatubo, 1991

In contrast with the Krakatau eruption, which illustrates the helplessness of past societies in the face of severe volcanic hazards, the Pinatubo eruption shows that even poor and developing states now have the technological and administrative capacity to minimize loss of life to volcanic eruptions. That being said, the government of the Philippines could not prevent the eruption from precipitating a cultural crisis among the Pinatubo Aetas, an Indigenous minority group that lived near or upon Mount Pinatubo before the eruption. Indeed, the main tool of the government to save lives, evacuations, was an important factor in turning the world of the Aetas upside-down.

Unlike Krakatau, which was a handsome conical peak before it was shattered by the 1883 eruption, Mount Pinatubo in 1991 was a nondescript and low-lying mountain in the Zambales range, barely distinguishable from the surrounding peaks. But their differences were only skin deep; like Krakatau, Pinatubo was a stratovolcano, built from successive layers of lava and pyroclastic materials, and fueled by subduction, which supercharges its magma with an explosive mix of water and dissolved gases. Pinatubo is only one of 200 known volcanoes in the Philippines, 16 of which were considered active at the time of the Pinatubo eruption, and 10 of which erupted during the twentieth century. The abundant volcanism of the Philippines owes its origin to the fact that the Philippine Mobile Belt mini-plate, which underlies most of the Philippine Archipelago, is experiencing subduction from both sides, as it is being squeezed between the Philippine Sea plate to the east and the Eurasian plate to the west.

Three main groups called the vicinity of Pinatubo their home prior to the 1991 eruption. One was servicemen of Clark Air Force Base, one of America's largest overseas military facilities. The area also hosted a large ethnic Filipino population, especially in the major urban center of Angeles City, which lay just 25 kilometers from Mount Pinatubo and less than 10 kilometers from Clark AFB. Pinatubo's closest neighbors, however, were the Pinatubo Aetas, a local branch of the Aeta ethnic minority. The Aetas, who are near genetic relatives of New Guinean highlanders and more distant relatives of the Australian Aboriginals, originally occupied the entirety of the Philippines. When Austronesian farmers from Taiwan began to migrate into the island chain in around 1500–1000 BCE, they gradually displaced Aeta hunter-gatherers, who lost most of their land as well as their pre-contact languages. As a result, today most Aetas occupy areas in the Philippines that are marginal for agriculture, including the rugged and densely forested Sierra Madre Mountains on Luzon island's northeastern shore – and more importantly to the present study, the slopes of Mount Pinatubo.

According to Japanese ethnographer Hiromu Shimizu (1989), who conducted research on the Pinatubo Aetas in the years before the eruption, the Pinatubo Aetas originally inhabited the fertile lowland terrain that now hosts Clark AFB and Angeles City but were displaced to the slopes of Pinatubo in the eighteenth and nineteenth centuries by planters of sugar cane and rice. This was not a bloodless process – well into the twentieth century, the lowlander Filipinos of the Angeles City regarded killing an Aeta as a sign of "machismo," and Aeta slaves were a common article of commerce. Still, while the Aetas might have moved up the mountain involuntarily, by the time of the eruption they had made Pinatubo their home, surviving there through a combination of animal husbandry, slash-and-burn agriculture, fishing in Pinatubo's streams, and hunting and gathering wild foods. The Pinatubo Aetas were so skilled in surviving in their environment, in fact, that the American military employed them to teach jungle survival skills to U.S. airmen during the Vietnam War era. Before the eruption, contact between the Aetas and lowland Filipinos was rare, and sometimes hostile, with conflicts revolving around land disputes between the Aeta and lowlander farmers.

In addition to providing their sustenance, Mount Pinatubo had become the central pillar of Aeta culture. The Aetas believed Pinatubo was the abode of Apo Namalyari, the supreme and creator god of the Aetas. The mountain was also the resting place of the dead, and prayers and offerings were customarily made in the mountain's direction. In the absence of the written world, the mountain itself was the record of Aeta history; as Shimizu puts it in his 1989 monograph, the Aetas "still see the river, mountains, caves, rock formation, streams and other landmarks where the most important events in their history happened … their past continues to be present through the images they see in their environment."

Within a decade of Shimizu's field work, however, the Aeta environment would be changed beyond recognition by Pinatubo's 1991 eruption. Luckily, Pinatubo telegraphed its explosive intentions ahead of time, first with earthquakes, which began in July of 1990, and then with increased sulfur dioxide emissions in early 1991. By June, molten rock was spotted on the volcano's surface, and the first small-scale pyroclastic flows began to tumble down the mountain. This provoked the Philippine authorities to evacuate everyone living within 20 kilometers of Pinatubo, which included many Aetas. Given the centrality of the mountain to their worldview, however, many Aeta were hesitant to leave. Most did not believe that Apo Namalyari would hurt his people, and in any case they thought his evident anger was directed, not at the Aetas, but at lowlander Filipinos who had recently drilled into Pinatubo's side to construct geothermal wells!

However, by early June, worsening Mount Pinatubo explosions forced the issue (Figure 4.2), compelling the movement of 25,000 Pinatubo Aetas to refugee camps. And not a moment too soon: on June 15, Alwyn Scarth (1999) tells us, Pinatubo launched "great cauliflower columns of ash, gas, and steam … up to 34km skywards." It also belched out as much as 10 cubic kilometers of pyroclastic materials and pulverized rock, enough material (according to Eddee Castro, 1991) to fill "71 million 10-wheeler trucks," lowering the height of the mountain by over 200 meters. Although the exact number is unknown, some Pinatubo Aetas apparently remained on the mountain, and all were killed by the June 15 explosion. However, among the Aetas and Filipino lowlanders alike, the overall mortality rate to the eruption itself was quite low. Due to early warning and effective

Figure 4.2 The June 12th eruption of Mount Pinatubo seen from Clark AFB
Source: U.S. Geological Survey Photograph taken by Richard P. Hoblitt.

evacuation, the eruption itself is thought to have killed only 200–300 people, a far cry from Karakatau's quintuple-digit death toll.

Although the refugee camps sheltered the Aetas from Pinatubo's pyroclastic flows and ashfall, most Aetas found life in the camps difficult. The hastily-built camps suffered from overcrowding, inadequate sanitation, and a lack of clean drinking water. These factors, combined with the fact that the Aetas were almost entirely unvaccinated, turned these camps into perfect stalking-grounds for disease. In the first 12 weeks after the eruption, health care workers reported that measles, respiratory diseases, and dysentery had killed 349 people in the refugee camps, and by the end of the first year this number had climbed to 537, 94% of whom were Aetas. Scarth speculates that the number of Aetas who died in the refugee camps may have actually surpassed the number of Aetas directly killed by the volcanic hazards of Pinatubo, though of course the death toll would have been vastly higher if most Aetas had not heeded the call for evacuation.

Life in the camps posed other problems for the Aetas as well. Aetas from the southern slopes of Pinatubo were taken to refugee camps that hosted both Aetas and lowlander Filipinos, triggering resource conflicts between the two groups, and there is some evidence that Filipino refugees received preferential treatment over the Aetas in the camps. Food was also an issue; in all the camps, the Aetas were forced to adapt to a new diet, including canned foods, which many Aeta did not understand. Worse yet, the destruction of their mountain meant that the Aetas were forced to remain in these camps for an extended period, but there was little good farmland near the camps, and the best such land was already owned by lowlander Filipinos. This contributed to further conflicts between the Filipinos and the Aeta, for whom the principle of private land ownership was an alien concept. The Aetas also had problems adapting to the rigors of full-time farming, since relying on just a few types of crops made them feel vulnerable, as they were used to consuming a wide range of seasonal foods while on Pinatubo. What is more, the discipline and long-term planning required by agriculture was hard for many Aetas to accept, since their previous hunter-gatherer lifestyle had favored opportunism and the flexible exploitation of immediately available windfall resources.

Most Aetas intended to return to Mount Pinatubo as quickly as possible after the eruption, but nature had other ideas. Pyroclastic flows had destroyed much of Pinatubo's vegetation, and while recovery began almost immediately in some slopes where local topography provided some protection from the volcanic hazards, scientists estimated that full regrowth of Pinatubo's formerly dense mountain forests might take 50 years. Worse yet, Pinatubo's slopes were coated with thick layers of pyroclastic materials, which killed off most of the local fish. These pyroclasts also formed the raw material for lahars, which repeatedly surged down Pinatubo's watercourses after the 1991 eruption whenever this unconsolidated volcanic rock was loosened by monsoon rains. These lahars convinced the government to construct the so-called "Megadike" at a cost of $30 billion in 1996 to restrain the lahars along the Pasig-Potrero River basin, which drains the eastern slopes of Pinatubo and flows by the outskirts of Angeles City. However, the dike suffered an enormous breach during the first rainy season after its construction, requiring costly repairs and signaling that lahars would continue to be a problem in the region for years to come.

In the last two decades, as the forests on the mountain slopes began to recover from the eruption, many Aetas have gone back to the slopes of Pinatubo. However, the Aetas who returned to the mountain were not the Aetas who had left it. According to Jean-Christophe Gaillard (2006), long exposure to lowlander Filipinos in the refugee camps had changed the Aetas, politically, culturally, and economically. The Aetas have abandoned some of their former egalitarianism in favor of stronger tribal chiefs, who serve as needed intermediaries with lowlander Filipino populations and the Philippine state. They are far more concentrated into larger, more permanent, and more agriculturally-based settlements, which lie farther down Pinatubo's slopes, facilitating closer interaction with lowlander Filipinos. Aetas are also far more likely than before to attend Filipino schools – their literacy rate has risen from 4 % to 30% since the Pinatubo eruption – and marriage between the Aetas and lowlander Filipinos is becoming increasingly common. Religiously, many Aetas have adopted aspects of Christianity, though without fully abandoning their own religious beliefs; the traditional ancestral spirits worshiped by the Aeta, for example, are now seen as manifestations of the "holy spirit" of the Christian trinity. More generally, the Aetas have incorporated more elements of modern western culture into their lifeways, including prepackaged foods, fast food, alcohol, and modern clothing, and building materials.

So did the Pinatubo eruption trigger a cultural crisis for the Aetas? According to Gaillard, it did not: the Aetas demonstrated considerable resilience in the face of both the disaster and lowland cultural influences, which has "underlined the flexibility and durability of Aeta society in the face of natural hazards." That being said, while the Aeta people seem to be doing well in the aftermath of the Pinatubo disaster, Aeta culture did not survive unscathed, and the Aetas now are far more assimilated into lowlander Filipino lifeways than before. Even on this front, however, there is hope for the survival of Aeta distinctiveness. According to Stefan Seitz (1998), contact with lowlanders in the refugee camps had the paradoxical effect of reinforcing Aeta cultural identity and their sense of separateness from the Filipino majority. What is more, some modern Aeta groups have found ways to monetize their cultural differences, for example, by participating in ecotourism trips to the slopes of Pinatubo, or manufacturing cultural artifacts for sale to visiting tourists. Whether this will be enough in the long run to preserve a distinct Aeta cultural identity in the face of accelerated post-Pinatubo cultural assimilation, however, remains to be seen.

Case Studies in Volcanism and Climate Change

Direct volcanic impacts due to physical hazards, therefore, have left an important mark on past societies. However, the indirect effects of earthquakes, especially short-term global cooling due to sulfate emissions, have been even more consequential over the course of human history.

If recent scientific studies are correct, in fact, the record of one of the most catastrophic volcanic eruptions in the Earth's history can be read in the modern human genome. One phenomenon that has long puzzled geneticists is the fact that, approximately 70,000 years ago, the human population seems to have dropped to only a few thousand individuals, creating a genetic bottleneck and greatly reducing human genetic diversity.

Similar bottlenecks dating to the same time period have been discovered in other species as well, including East African chimpanzees, Borneo orangutans, and others. Some scholars, most notably anthropologist Stanley H. Ambrose (1998), have argued that this bottleneck is best explained by mass mortality episodes triggered by volcanically-induced climate change, specifically the multi-year volcanic winter that might have followed the eruption of Mount Toba in Sumatra about 74,000 years ago. Geological evidence suggests that Toba exploded with approximately three thousand times the force of the Mt. Saint Helens eruption, and it may have expelled up to 6,000 square kilometers of pyroclastic materials – 600 times the volume of the Pinatubo eruption, enough to bury 1% of the planet's surface beneath 10 centimeters or more of ash. Based on ice core analysis, it also released an enormous volume of sulfates, perhaps 100–300 times the amount released by Pinatubo. As a result, some climate scientists claim, the Toba eruption may have caused global temperatures to drop by 3°C or more for six to ten years, killing most of mankind in the process. However, the "Toba Catastrophe Theory," as it is called, is not fully accepted by all climate scientists.

The Great Climate Event of 536 CE

Equally controversial, though better documented, is the volcanic winter that began in 536. In that year, according to the Byzantine historian Procopius, "the sun gave forth its light without brightness, like the moon." In Italy, the Byzantine civil servant Cassiodorus Senator worried about that year's "blue-colored" sun, which shone its feeble rays over "a winter without storms, spring without mildness, summer without heat." On the other side of the Eurasian continent, in China, observers complained of a persistent dry fog, and chroniclers in northern China recorded that "frost and snow in July and August killed the seeding crop, causing a major famine the following autumn." Unusually cold and dry weather was reported in China until at least 538, and in one kingdom north of the Yellow River, 70–80% of the population reportedly died of famine. The disaster even touched the distant Americas, as evidenced by the Tikal "hiatus," the complete absence of any carved stone monuments in the regionally dominant Mayan city of Tikal between the years of 534 to 613 CE. Nor is our evidence limited to the written record alone. Significant deposits of sulfates and volcanic ash have been detected in ice cores dated to around 536 CE, and in the 15 years after 536, Irish oak trees grew at the slowest rate than they had in the entire 6,000-year period that can be reconstructed using tree ring sequences.

All the signs point to a major volcanic eruption or eruptions in 536 CE, therefore, but even today it is not entirely clear which volcanoes were involved. In his 1999 book on the 536 CE disaster, which we will examine below, David Keys speculated that the culprit was a proto-Krakatau volcano, though subsequent evidence has proven Krakatau's innocence. Other scholars have blamed the Ilopango Volcano in El Salvador, the Rabaul Volcano of Papua New Guinea, or an unknown Icelandic volcano. A study by M. Sigl et al. (2015), in turn, has argued that at least two different eruptions were responsible for the event, including a North American eruption in 536 and second event in the tropics in 539–540, which some scholars believe was the eruption of El Chichón volcano in southern Mexico, as we will see below. This double whammy of climate events, the authors argue, released enough sulfates into the atmosphere to make the decade from 536–545 CE the coldest decade in 2,500 years. More recent studies by Ulf Büntgen et al. (2016) and Peter N. Peregrine (2020) have argued that the global volcanic winter that began in 536 may have been triggered by *three* volcanic eruptions – in 536, 540, and 547. Peregrine contends that this "Late Antique Little Ice Age" (LALIA) did not end until 560 CE, while Büntgen et al. would push its end date all the way to 660. As we will see in Chapter 5, some of this confusion about when the LALIA ended arises from the fact that it overlapped with the "Dark Ages Cold Period," a cold snap probably caused, not by volcanism, but by a reduction in radiative forcing.

Uncertainty about the exact origins and time span of the event, however, has not prevented some authors from making sweeping claims about the world-shaking consequences of the 536 climate disaster. In *Catastrophe*, for example, journalist David Keys (1999) implicated the 536 climate disaster in a number of major historical events, including the outbreak of bubonic plague, the start of Slavic invasions of the Byzantine Empire, the rise of the Turks, the collapse of the Marib dam in Arabia and the rise of Islam, the adoption of Buddhism in China, Korea, and Japan, the re-unification of China after a protracted period of warring states, and the fall of both the Mayan city of Tikal in Central America and the Moche civilization in Peru. Some of these events (such as the rise of the Islamic imperial state) occurred long after 536, but Keys insists they are connected to 536 via the near-simultaneous outbreak of the first global pandemic of bubonic plague. This disease jumped to humans, Keys contends, due to climate-change-induced alterations in the behavior of rodents, who are the natural reservoir of *Y. pestis* bacteria. Thus, Keys ascribes the success of the Arab conquest over the Byzantines in the early seventh century not only to the Byzantine loss of much of the Balkans to the Slavs in the sixth century, but also to the extreme financial problems Byzantium faced after the repeated plague epidemics since 541, which left the Empire with a much-depleted number of taxable citizens.

Although Key's work has not enjoyed mainstream acceptance, other scholars have made similar arguments about the LALIA's impact on history. Büntgen et al. (2016) argue that proxy evidence from ice cores, tree rings, lake sediments, and other sources strongly supports the notion that the LALIA shortened the pasture-growing season in the Central Asian steppes. The resulting turmoil, they argue, probably played an important role in impelling the Turkish invasions into China and the proto-Slavic and Lombard invasions into the Roman Empire, further weakening the Byzantine state. Interestingly, they also outdo Keys by drawing a more direct connection between the Arab conquests and the LALIA, speculating that reduced evapotranspiration in the desert due to cooler temperatures may have promoted scrub growth in both the Arabian Peninsula and the surrounding deserts. This would have created excellent conditions for camel breeding, and the resulting herds "may have facilitated transportation of the Arab armies and their supplies during the substantial conquests of the seventh century." Although Büntgen et al. do not make this point, I would add that Khalid b. Al-Walid's famous 634 CE march across the Syrian Desert to reinforce an Arab offensive against the Byzantines is a perfect illustration of how the growth of scrub vegetation might have assisted the Arab conquests. As I noted in a 2012 paper, Khalid's march took place during the spring season when desert vegetation would have been green and flourishing, and thanks to their lactose tolerance, Khalid's army could have derived hydration even in the waterless desert from the milk of their camels (Reilly, 2012).

In addition to facilitating the Arab conquests, the LALIA and the resulting plague epidemic may also help explain the conquest of England by the Anglo-Saxons. In a fascinating chapter, anthropologist Elizabeth Jones (2016) points out that the "dry fog" of the 536 would have disproportionally impacted the Anglo-Saxons, since a cold snap would have been especially disastrous in their Scandinavian homeland, which already had a short growing season even before factoring in a volcanic winter. As a result, the invading Anglo-Saxons could be likened to climate refugees, albeit refugees with arms. As for the Britons, she argues, they were far more impacted by the post-536 outbreaks of plague than were the Anglo-Saxons, thanks to their close trade connections to the European mainland and their far more urbanized population. Devastated by epidemics, the Britons were in no position to hold off the Anglo-Saxons during the 540s, and archeological evidence of unburied but unwounded corpses lying in the streets of Briton towns suggests that some plague-stricken Anglo-Saxon urban settlements were abandoned by their inhabitants even before the Anglo-Saxons arrived on the scene. Jones also speculates that disease helps to explain why the Anglo-Saxons refused to settle in the cities they conquered, since they saw them as haunted tombs of the dead rather than spoils of war. An early Anglo-Saxon poem describes one of these abandoned towns as depopulated by "days of pestilence," which "swept away all the bravery of men; their fortresses became waste places; the city fell to ruin. The multitudes who might have built it anew lay dead on the earth."

Jones even argues that the climate impact of the 536 event helps to explain why the legendary King Arthur's death is generally dated to 537–539. In Celtic culture, kings were semi-sacred creatures, and the health of a king was tied to the well-being of the kingdom. As Jones puts it, "a healthy, just and victorious king would be reflected in a fertile land and healthy subjects," while an "injured or maimed king ... would be reflected in a blighted, barren landscape." In support of this, she notes that the wizard Merlin is made to say in the legends of King Arthur that, due to Arthur's unhealed wounds, "the very stars would wither the standing fields of grain and forbid the rain to fall from heaven." As Jones points out, Merlin's lament about the state of Britain is "not ... a bad description of the climatic event and dry fog of A. D. 536." The fact that Arthur's death is dated to the same period of the 536 climate event, therefore, is not coincidental, but may represent an attempt to construct a cultural explanation for the 536 event by linking the calamitous weather to the tragic downfall of a legendary king.

Recent scholarship also gives some support to Key's contention that the 536 event may have helped trigger the decline of the Mayan city-state of Tikal, though it has complicated Key's simplistic one-eruption model for the 536 climatic catastrophe. According to a paper by Kees Nooren et al. (2017), Tikal and the other cities of the Mayan lowlands were probably already suffering from the 536 disaster when the nearby El Chichón volcano erupted violently in 540. This eruption probably exacerbated the ongoing volcanic cooling, but more importantly, it did direct damage to Tikal itself through ash fall and subsequent lahars, undermining Tikal's agricultural production. This same ashfall, however, seems to have done relatively little damage to Tikal's rival Caracol, which joined forces with Calakmul to defeat Tikal during the so-called "Star War" of 562 CE. Tikal's king was captured and sacrificed to the gods, and Tikal's public monuments were mutilated by the victorious coalition. It would take Tikal until the late seventh century to recover from this crushing defeat, but if Nooren et al. are correct, the real root of Tikal's problems was the environmental impact of the El Chichón volcano.

While studies like those of Jones, and Nooren et al., have examined the 536 event's impacts on specific locations, anthropologist Peter Peregrine's (2020) paper on the "Late Antique Little Ice Age" takes a broader view. Peregrine uses data from 20 different geographically scattered case studies, ranging from Byzantine Egypt to Early Classic Tikal in the Mayan Lowlands to the Gupta Empire in India, in order to conduct a "naturalistic quasi-experiment" on whether or not the LALIA led to dramatic change, in terms of changes in population size, frequency of conflict, religious practices, and other factors. The short answer is yes, it did: "dramatic climatic change," he finds, "is empirically linked to dramatic social change." That being said, he notes that there was considerable variation in how societies reacted to climate challenges. While some societies experienced significant and sometimes destructive changes during the LALIA period, others weathered the crisis quite well. This variety of outcomes, Peregrine believes, is best explained by each site's own "environmental, technological, and sociopolitical vulnerabilities," which "must be addressed in any consideration of climate-related social change." Peregrine's comparative study of the 536 event, therefore, serves as a reminder of how important it is to consider hazards in the context of specific local vulnerabilities, which mediate how the disaster (if there is one) will ultimately unfold.

Laki, 1783–1784

While events such as the Tikal "hiatus" of 534–613 CE epitomize the ability of volcanic events to alter history through their direct impacts, Laki's eruption in 1783–1784 reminds us that volcanic eruptions can also kill indirectly, through chemistry rather than physics. As we saw at the start of the chapter, Laki's eruption was not explosive, and the main physical hazard it presented, its lava flows, traveled at less than a quarter a kilometer per hour and thus were more of a danger to immobile buildings and farmland rather than livestock and people. The real threat Laki posed was at the molecular level, since Laki's abundant lava flows "de-gassed" 7 million tons of chlorine and 15 million tons of fluorine into the atmosphere. In addition, Laki released as much as 122 tons of sulfur dioxide into the atmosphere, the equivalent of 12,000 years' worth of sulfur emissions from a modern coal-fired power plant, and more sulfur than any eruption had released in the past 1,000 years. These emissions would wreak such havoc on the global climate that some scholars have dubbed Laki the deadliest natural disaster in human history.

To understand Laki's eruption, we first have to understand the volcanic forces that have created Iceland. Like Japan, Iceland was built by volcanism, but unlike Japan, Iceland's volcanoes consist, not of elegant conical stratovolcanoes, but rather of chains of "splatter cones," where lava flows freely and mostly non-explosively at the world's surface. The difference arises from plate tectonics. While Japan's volcanoes are fueled mainly by subduction along a thrust fault, Iceland's volcanoes lie atop "normal" faults, where two plates are moving away from each other, creating space for magma to seep to the surface like blood from a gash in the skin. Volcanoes of this sort are typically found under the ocean, but Iceland is one of only a few places on Earth where this process is occurring on dry land, perhaps because Iceland's volcanism arises from a "hot spot" plume in the mantle in addition to the forces of plate tectonics. Differences between Iceland and Japan's volcanos are not just a matter of style, but a matter of substance. While Japan's stratovolcanoes are rich with dissolved silicates, CO_2 and H_2O, Iceland's pyroclastic materials feature high levels of iron, to the point that observers in Venice at the time of Laki eruption discovered to their surprise that Laki's ash was attracted to magnets.

The most dangerous chemicals emitted by Laki, however, were its chlorine, fluorine, and sulfates. At ground zero of the eruption, in Iceland,

Laki's emissions led to the deaths of half of the island's livestock and a fifth of the island's population, mostly through fluorine poisoning, though starvation played a role as well. Elsewhere in Western Europe, Laki's sulfates were likely responsible for the "black fever" nose and throat ailments that were widely reported in Europe during the summer of 1783. This black fever was almost certainly the result of the inhalation of Laki's gaseous emissions, especially sulfur dioxide, which turns into sulfuric acid in the human body and can be deadly at high concentrations. Contemporaries described this black fever as "nearly as fatal as the plague," and while this description is perhaps a touch overdramatic, England alone probably suffered 20,000 or more excess deaths in the summer of 1783. According to Laki experts Alaxandra Witze and Jeff Kanipe (2014), if an Icelandic eruption on the same scale were to occur today, the death toll would likely top 140,000.

Laki's "dry fog" of sulfates wreaked havoc, not just on Europe's lungs, but also on Europe's climate and weather. Observers throughout Europe described this fog in dramatic terms: English poet William Cowper complained in a letter written amidst Laki's haze that "we never see the sun but shorn of his beams, and the trees are scarce discernible at a mile's distance." When this feeble sun sets, Cowper noted, it descends "with the face of a hot red salamander," and it "rises [the next day] with the same complexion." In a more scientific vein, American scientist Benjamin Franklin, who was in Europe to negotiate peace at the end of the American Revolutionary War, wrote that during in the summer of 1783, "there existed a constant fog over all Europe ... this fog was of a permanent nature; it was dry, and the rays of the sun seemed to have little effect towards dissipating it." Franklin speculated correctly that this dry fog was the result of a volcanic eruption in Iceland, but in general, Europeans interpreted Laki through a religious lens, seeing in Laki's haze the wrath of God or a herald of Judgment Day.

Although sulfates generally trigger global cooling, the summer of 1783 was actually quite hot in northern and western Europe, and this heat was punctuated by frequent and terrifying thunderstorms that killed many people as well as some of the horses of French King Louis XVI's royal herd. It is unlikely that Laki's sulfates directly caused this effect; rather, it is more likely that Laki turned the heat up in parts of Europe by blocking normal atmospheric circulation patterns, allowing warmer air from the southwest to govern Europe's climate. In any case, when November came, this freak summer heat gave way to a bone-chilling volcanic winter that was 3°C colder than the 30-year average throughout most of the Northern Hemisphere. Worse yet, we should keep in mind that Europe's temperatures at the time were *already* colder than normal; as we will see in Chapter 5, Laki occurred in the middle of a multi-century cold snap called the Little Ice Age, which magnified Laki's effects. When this volcanic winter finally ended in April or even May of 1784, the thaw melted so much ice and snow that rivers throughout Europe experienced record floods, which washed away bridges and laid waste to riverside villages. In the US, so much ice was carried down the Mississippi during the spring thaw of 1784 thaw that an ice jam near New Orleans temporarily blocked the mouth of the Mississippi River. However, despite the severity of the winter of 1783–1784, there are no reports of mass famine or starvation in Europe or North America, in part because the cold only set in after the 1783 harvest had already been collected. The only exception to this rule was in Iceland itself, where an estimated 20–25% of the population died, largely due to fluorine poisoning and the decimation of livestock herds. Half a world away, Japan did suffer a major famine during this period, the Tenmei Famine. However, that famine is generally blamed on local causes, including the eruption of Japan's own Asama volcano and to fiscal mismanagement by the Tokugawa Shogunate.

Ironically, the people to suffer most from the Laki-induced cold snap may have been the inhabitants of the subtropics, particularly in Egypt and India. In normal years, the rapid summertime heating of the Eurasian land mass compared to the Indian Ocean helps to draw Indian Ocean moisture northwards, helping to fuel the annual cycle of monsoon rains. However, according to a study by Luke Oman et al. in 2006, Laki's chill led to a "decreased temperature gradient between Asia and Europe and the Indian and Atlantic oceans," weakening the Indian Ocean monsoon. As a result of this, Oman et al. argue, Egypt suffered from two successive Nile flood failures, in 1783 and again in 1784. Together, they triggered a famine which (in the words of the visiting French traveler, the Comte de Volney) "carried off, at Cairo, nearly as many as the plague; the streets, which before were full of beggars, now afforded not a single one: all had perished or deserted the city". Volney estimated that by January of 1785, Egypt had lost fully a sixth of its population to death or emigration.

Monsoon-dependent India also suffered a serious famine as well in 1783–1784, the Chalisa or "forty-year" famine, which mainly afflicted northern India. Since the hungry regions were under the control of a variety of local Indian princes rather than the British, our information about this famine is limited, but nineteenth-century historian F. C. Danvers (1877) argued that northern India was "visited with extraordinary drought" during this period, during which "the crop upon the ground had been scorched, and nearly destroyed." The death toll apparently ran well into the millions, though again, the event is little documented.

So is it true that the Laki eruption was the deadliest natural disaster in history? Perhaps, but in the case of Egypt and India, the Laki eruption's effects are difficult to distinguish from the workings of the El Niño/ Southern Oscillation (ENSO), a cyclical weather system that often triggers Indian Ocean monsoon failures when it is in the "El Niño" mode, as was the case from 1782–1784. In any case, a single harvest failure rarely leads to a famine, so the 1783–1784 famine would owe its origins as much to a weak 1782–1783 harvest as to any impact of Laki in 1783–1784. The fact that monsoon failures would be extraordinary common in India from the 1780s all the way until 1810, well after any cooling effect from Laki would have been long over, also suggests that ENSO might have been the main culprit for the Chalisa famine and that Laki might have played only a supporting role. We will return to the topic of ENSO's connection to the monsoon cycle in Chapter 5.

"Eighteen Hundred and Froze to Death": the Tambora Eruption

The most recent volcanic event to cause large-scale and deadly climate changes is the 1815 eruption of the Tambora stratovolcano of Sumbawa Island, which lies about 1,600 kilometers to the east of Krakatau in the same arc of volcanic peaks. Tambora's eruption went mostly unnoticed by the world, but this was due to its remoteness rather than the severity of its impacts. An estimated 88,000–100,000 people died due from the eruption in the immediate vicinity of Tambora, in fact, succumbing either to physics (the explosion itself, pyroclastic flows, or Tambora's 4-meter-high tsunami), to chemistry (especially fluorine poisoning), or to hunger, as Tambora's ash smothered nearby agricultural fields. As bad as Tambora's eruption was for local Indonesians, however, its global climate impacts may have been even worse. Thus, 1816, the year after the eruption, became known as the "year without a summer," and it precipitated what historian John D. Post (1977) has called the "last great subsistence crisis of the western world."

At the time that Tambora blew its top, releasing up to 25–30 megatons of sulfur dioxide into the atmosphere in the process, much of Europe was already in a vulnerable state. The continent had just emerged from a draining 25-year period of near-constant warfare, triggered by the French Revolution of 1789 and the subsequent rise of Napoleon Bonaparte. The final battle of the Napoleonic wars, Waterloo, occurred just two months after Tambora's main eruption. What is more, as in the days of Laki, Europe in 1815 was still feeling the chill of the Little Ice Age, which dropped temperatures in the Northern Hemisphere as much as 2°C below that of the previous period. Tambora's sulfates would reduce the world's temperature by a further .5°C or so, and this effect would persist for about three years. Therefore, 1816, in fact, proved to be the second coldest year in the Northern Hemisphere since 1400 CE, and in some places in the Eastern U.S. and Canada, significant snowfall was reported in the first weeks of June. As late as July 5, according to Henry Stommel and Elizabeth Stommel (1983), one town in Maine woke to an unheard-of sight – a layer of ice "thick as window glass" had formed on water bodies overnight. Throughout the Northern Hemisphere, there were record-setting delays in the start of plant growth, to the point that in many places the crops were still immature and unripe in November, when the first frosts of 1816 brought an end to the growing season.

As might be expected, the parts of Europe and North America most impacted by Tambora's volcanic winter were places that were marginal for agriculture even in normal years, such as lands in high latitudes or at high elevations. Iceland suffered yet another famine due to Tambora, as did parts of Ireland, Wales, England, and Germany. The economic suffering spawned by Tambora's chill seems to have been particularly severe in the mountains of Switzerland. In the Swiss town of Neuchâtel, for example, industrial workers were already suffering even before the Tambora eruption due to renewed British economic competition following the end of Napolean's embargo of British goods. During the "year without a summer," these impoverished workers were further victimized by a cruel combination of rising grain prices and falling demand for the manufactured goods they produced. Such was their poverty, one observer wrote, that

> three to four households with numerous children live crowded together in a single room, dressed in colorless rags, half naked, completely naked. Tables, chairs, bed-posts have long since disappeared [probably sold for food]. Their nourishment consists of roots, herbs, and some scrapings of yeast. People are dying from want of a drop of blue milk. Summer and winter they all lie in their rags on the hard floor, near the stove if old age or sickness affords them that privilege. Not a few resemble skeletons dragged from the grave, and the infant who should be a joy to the world emerges like a corpse from the mother's womb.

Much the same misery engulfed the northern parts of the United States, which like Europe, had just emerged from a damaging war, in this case the British-American War of 1812. Food prices, which had already risen precipitously due to British embargos during the war, spiked once again due to post-Tambora harvest failures, from a pre-Tambura low of 9 shillings per bushel of wheat to 18 shillings by 1817. Many Americans tried to counter these increased food prices by purchasing more mackerel from the market or setting up seines and nets in local rivers to catch these fish for themselves, earning 1816 the nickname the "mackerel year." What is more, the high grain prices were of little benefit to local farmers, who suffered meager crop yields; as far south as Connecticut in the United States, as much as three-quarters of the maize crop succumbed to the low temperatures. *Moby Dick* author Herman Melville noted that, during the lean Tambora years, New Bedford whaling ships were full of "green Vermont and New Hampshire country boys," who had been driven off their lands by the cold and were trying their luck in the warmth of the South Seas instead.

Other displaced populations tried to survive Tambora's chill through migration. In Switzerland, thousands of landless farmers and out-of-work craftsmen took to the streets, "flow[ing] hrough the towns and villages in droves," according to Tambora historian Wolfgang Behringer (2019), and "loudly crying out for the first necessities in life. The roads were filled with these unfortunates like armies on the march, market squares were their gathering places, entire streets were as if besieged." In addition, many farmers fled lands left infertile by Tambora for what they hoped would be greener pastures elsewhere. Out-migration from the German state of Württemberg, which lies just north of Switzerland, rose from just over 400 per year in pre-Tambora 1812/1813 to 17,500 in 1816/1817, and many of these former Württembergers were headed to Russia or the USA in search of farmland. In Vermont, New Hampshire, and Maine in the United States, individual families sold their hardscrabble farms and headed for western destinations like Indiana, Ohio, and Illinois. Joseph Smith, the eventual founder of Mormonism, was one such climate refugee – his family was forced out of Vermont by crop failures and eventually settled in western New York.

Still other Europeans reacted to Tambora with violence, especially in France, where lean Tambora harvests struck a society that had only meager stocks of stored grain due to wartime food requisitions. In Poitiers in the Loire Valley, for example, Stommel and Stommel (1983) note that food carts on the way to the markets had to be protected by "soldiers and gendarmes, who found themselves fighting on occasion as many as 2,000 hungry and enraged citizens." In another incident, a riot broke out Fauville, Normandy, after food stocks at the market declined from the usual 800–900 sacks of grain to only 50. Stone-throwing local citizens descended upon the market, drove away the soldiers, and seized what little grain there was to be had. Such scenes, Stommel and Stommel note, were reenacted "in little towns all over France" in 1817, until an adequate late-1817 harvest brought some improvement to food prices and supply.

Food "riots" of this sort, incidentally, were not random outbreaks of greed and brigandage, but rather manifestations of a deeply-held European "culture of famine" designed to protect communities in times of general dearth. According to an influential article by social historian E. P. Thompson in 1971, traditional European societies adhered to a set of beliefs that Thompson dubbed the "moral economy of the poor": the principle that in times of scarcity no man should profit from another man's starvation, that food prices should remain at customary levels even if stocks were low, and most importantly, that food should not leave any settlement suffering from scarcity. These beliefs were so widespread that, according to Stommel and Stommel, some local officials even took the side of the "rioters" during these Tambora-era struggles over provisions, since they shared the popular belief that "harvest of their region belonged to its inhabitants and exportation was a violation of their security." In any case, Wolfgang Behringer (2019) claims that the elite fears provoked by these food "riots" may have joined with more generalized anti-Revolutionary sentiment to inform the reactionary conservativism of the post-1815 era, as epitomized by the draconian 1819 Carlsbad Decrees in Germany and the Peterloo Massacre in Britain during the same year.

The mob violence of the Tambora era may have also played a role in one of the period's more shameful episodes, the xenophobic scapegoating of German Jews. In the years after the Tambora eruption, soaring grain prices were increasingly blamed on the "corn Jews," who were said to be driving up demand artificially through speculation and hoarding. Antisemitic anger came to a head in 1817–1818 with the "Hep-Hep" riots, which started in Austria and then spread widely through Germany (Figure 4.3). Popular mobs injured many Jews, looted, and/or torched their businesses and residences, and threatened them with massacre. As Wolfgang Behringer points out, most of the Hep-Hep rioters were people "on the losing side of the famine" – they were those who Tambora had left hungry, indebted, and unemployed. He also notes that the Hep-Hep riots had some specific local causes. in Wurzburg, for example, the rioters seem to have been motivated by resentment against Bavarian domination as much as rumored Jewish profiteering.

While culture and history certainly played a role in the Hep-Hep riots, especially insofar as it funneled popular anger in the direction of the Jews, the xenophobia of the Hep-Hep rioters may also have been rooted in the behavioral immune system, which becomes more sensitive to disease cues during periods of hunger and malnutrition. And disease cues were plentiful in the

Figure 4.3 Print of the Hep-Hep antisemitic riots of Bavaria, 1819
Source: Objektdatenbank, https://datenbank.museum-kassel.de/314137

aftermath of Tambora. While there was no recurrence of Laki's "black fever" in 1815, the malnutrition and human migrations of the Tambora era contributed to a massive epidemic of typhus, a disease spread by body lice that thrives in conditions of overcrowding and squalor. Typhus afflicted an estimated 2.5 million Europeans between 1816 and 1819 and may have killed 65,000 or more. What is more, during the Tambora crisis, parts of Italy suffered from epidemics of pellagra, a potentially deadly nutritional disease often caused by a maize-rich but vitamin-poor diet. These and other diseases combined with malnutrition sharply increased the mortality rate throughout Europe in 1817, which rose 22% in Switzerland, 25% in Tuscany, and 40% in the French region of Lorraine.

However, the epicenter of suffering in Europe during the Tambora climate crisis arguably occurred in Ireland, where famine and subsequent disease outbreaks combined to kill 140,000 or more between 1816–1817. As was the case throughout Europe, the events of the previous two decades before 1815 had greatly increased Ireland's vulnerability to the hazards posed by the Tambora crisis. Up until 1815, Ireland had enjoyed relative prosperity, in large part because of high wartime British demands for Irish grains and linens. Consequently, the population of Ireland had soared from just over four million in 1780 to just over six million by the eve of the Tambora eruption, a rate of growth made possible in part by the high caloric yield of Ireland's staple subsistence crop: the potato. However, the end of the Napoleonic Wars combined with the trauma of the "year without a summer" brought the Irish economy crashing back down to earth. The cessation of hostilities after Waterloo lowered food prices and opened Irish industries to foreign competition, throwing many out of work. What is more, as 1815 gave way to 1816, Ireland was struck by unseasonal cold as well as excessive rains, in some cases double the normal amounts. One Irish observer announced that the summer of 1816 was without parallel in the "memory of man … wet, cold, and in every respect uncongenial to the growth of maturation of the fruits of the earth." Overirrigated in Ireland's waterlogged soils, Ireland's potato crop succumbed to a variety of fungal and bacterial infections, such as "blackleg, soft rot, white mold, and powdery scab."

The resulting hunger, according to Gillen D'Arcy Wood (2014), drove the Irish to desperate measures in 1816 and early 1817:

> Starving people roamed the woods in search of "ramps" – a wild onion considered disgusting in ordinary times. Girls shaved their heads and sold their hair to peddlers for a pittance, while families bled their half-scarved cattle, feeding on the blood mixed with a little barley – truly a soup of the damned.

Making matters worse, the incessant rains meant the peat deposits the Irish normally relied upon to heat their homes were waterlogged and incombustible. As a result, whole families of the Irish poor huddled together in their beds for warmth, creating an opportunity for louse-borne typhus. One Irish doctor reported that, when he made house calls, he "frequently found all the members of the family laid in the same bed with a patient labouring under fever, owing to their having but one or two blankets." What is more, Ireland's typhus victims elicited little sympathy from the British and Irish urban elites, who, "in the grip of their paranoia of contagion," regarded starving rural refugees as "vagrants and beggars," who had "emerged from receptacles of disease, and spread themselves in various directions." The *Dublin Evening Post* demonized these hungry famine victims, claiming that even a beggar's "touch … nay, even the air which is about him, are pestiferous." As a result, rather than accept the starving Irish into their towns, urban authorities constructed "fever huts" outside urban lines, where the Irish poor were left to die. It would be hard to imagine a more obvious historical example of how the activation of the behavioral immune system by disease cues can trigger dehumanization of out-groups and exclusionary behavior.

While the impacts of the Tambora crisis are best known in Europe and North America, they were by no means limited to those regions. As was the case after the Laki eruption, Tambora's volcanic winter inhibited Indian monsoon by reducing the temperature differential between land and sea. While Egypt did not suffer from famine in 1815–1817, which suggests that the East African monsoon was not seriously impacted by Tambora, India suffered numerous harvest failures in the same period due to weak monsoons. However, India's monsoon failures were overshadowed in this case by an unprecedented outbreak of cholera, a disease that had formerly been seasonal and endemic in Bengal. In the aftermath of Tambora, cholera exploded into a fast-spreading epidemic infection, which would go on a worldwide killing spree in the decades that followed. To what degree this is Tambora's fault is not entirely clear. Stommel and Stommel (1983) certainly think Tambora was to blame, and Tambora expert Bernice de Jong Boers (1995) agrees that the "year without a summer" might have played an indirect role in triggering the first cholera epidemic, noting that "abnormal weather, the crop failures and subsequent famine" during the Tambora crisis might have "reduced resistance to the disease." However, this is just speculation.

There is also evidence that the southern Chinese frontier province of Yunnan suffered horribly from famines induced by monsoon failure during this period, with significant long-term consequences. Although Yunnan is on average about 2,000 meters above sea level, it nonetheless is one of China's balmier provinces, thanks to the warmth of the Pacific Ocean monsoon, which typically blows through Yunnan from June to October. Yunnan's farmers relied on the monsoon's heat and moisture to grow rice, the region's stable crop, and such was their expertise that local farmers could coax two or even three crops a year out of Yunnan's rice paddies, which extended in terraces well up the slopes of Yunnan's steep mountainsides. As a result of these high rice yields combined with a deliberate Qing dynasty settlement program of ethnic Han Chinese, Wood tells us, Yunnan's population had jumped from just three million in 1750 to 20 million in 1820, a feat only possible thanks to an "unwritten contract between Qing ambition and a benign climate."

But this contract was broken in 1815–1818. The monsoons failed, and were replaced by cold, desiccating northern winds, dropping summer temperatures by up to 3°C and reducing the rice harvest by up to two-thirds. Snow fell in Yunnan in the summer of 1817, an unheard-of circumstance. Normally, the Qing state retained enough of a grain stock in its "ever-normal granaries" to get its subjects through a bad harvest, but three bad years in succession had emptied the region's food reserves, and the remoteness of Yunnan from the rest of China complicated efforts to shift food stocks into famine-stricken Yunnan. As a result, Wood claims, by 1817, "famished corpses lay unmourned on the roads" in Yunnan, while "mothers sold their children or killed them out of mercy; and human skeletons wandered the fields, feeding on white clay." Many others turned to emigration for survival, much like their contemporary counterparts in Switzerland, Württemberg, and Maine. In addition, many in Yunnan who survived the crisis would abandon rice cultivation for opium poppies, a hardier and more valuable crop, but one that contributed to a national Chinese crisis of addiction and dependency. The Chinese would later suppress Yunnan's poppy-growing industry but it would only migrate southwards to modern-day Myanmar, Thailand, and Laos, the "golden triangle" of the modern-day opium trade. As a result, Wood argues that Tambora can be at least tangentially connected to the "first emergence of the modern international illicit drug trade."

As the multiple case studies of this chapter have illustrated, the ability of volcanism to change history indirectly through the chemistry of climate change is just as significant, if not more so, as its influence on history via direct physical hazards. However, volcanoes are far from the only factors capable of altering our climate. As John L. Brooke (2014) points out in his influential history of climate change, scholars in the field are divided into two warring camps, the "geologist-plumbers" who try to relate extinctions and other major climate events to tectonic forces within the planet, and the "astronomer-star gazers" who argue instead for the importance of extraterrestrial impacts, such as the asteroid or comet whose impact 35 million years ago gouged out Chesapeake Bay. In addition, Brooke notes, a growling literature connects climate changes to variations in solar energy output, leading to climate fluctuations on the scale of years, decades, and even centuries. These oscillations, and their impact on human history, will be the subject of Chapter 5.

Further Reading

Chester, David K. "Volcanoes, Society, and Culture." In Joan Marti and Gerald G. J. Ernst (Eds.), *Volcanoes and the Environment*. Cambridge: Cambridge University Press, 2005, pp. 404–439.

Dahlin, Bruce H. and Arlen F. Chase. "A Tale of Three Cities: Effects of the AD 536 Event in the Lowland Maya Heartland." In Gyles Iannone (Ed.), *The Great Maya Droughts in Cultural Context*. Boulder, CO: University Press of Colorado, 2014, pp. 127–155.

Damodaran, Vinita *et al.* "The 1780s: Global Climate Anomalies, Floods, Droughts, and Famines." In Sam White, Christian Pfister, and Franz Mauelshagen (Eds.), *The Palgrave Handbook of Climate History*. New York: Palgrave Macmillan, 2018, pp. 517–550.

De Rose, Ronald Charles *et al.* "Land Cover Change on Mt. Pinatubo, the Philippines, Monitored using ASTER VNIR." *International Journal of Remote Sensing*, vol. 32, no. 24 (2011). doi:10.1080/01431161.2011.554452.

Espíndola, J. M., J. L. Macías, R. I. Tilling, and M. F. Sheridan. "Volcanic History of El Chichón Volcano (Chiapas, Mexico) During the Holocene, and its Impact on Human Activity." *Bulletin of Volcanology*, vol. 62 (2000), pp. 90–104.

Grattan, John and Mark Brayshay. "An Amazing and Portentous Summer: Environmental and Social Responses in Britain to the 1783 Eruption of an Iceland Volcano." *The Geographical Journal*, vol. 161, no. 2 (1995), pp. 125–134.

Grattan, John, M. Durand, and S. Taylor. "Illness and Elevated Human Mortality in Europe Coincident with the Laki Fissure Eruption." In C. Oppenheimer, D. M. Pyle, and J. Barclay (Eds.), *Volcanic Degassing*. London: Geological Society, Special Publications, vol. 213 (2003), pp. 401–414.

Gunn, Joel D. (Ed.) *The Years Without Summer: Tracing A.D. 536 and Its Aftermath*. Oxford: BAR Publishing, 2016.

Houston, Margaret Snow. "Chinese Climate, History, and State Stability in A.D. 536." In Joel D. Gunn (Ed.), *The Years Without Summer*. Oxford: BAR Publishing, 2016, pp. 71–77.

Issar, Arie S. and Mattanyah Zohar. *Climate Change: Environment and History of the Near East*. 2nd ed. New York: Springer, 2007.

Luterbacher, J. and C. Pfister. "The Year Without a Summer." *Nature Geoscience*, vol. 8 (2015), pp. 246–248.

Lutgens, Frederick K. and Edward J. Tarbuck. *Foundations of Earth Science*, 7th ed. New York: Pearson, 2014.

Oppenheimer, Clive. *Eruptions that Shook the World*. New York: Cambridge University Press, 2011.

Prager, Ellen. *Furious Earth: The Science and Nature of Earthquakes, Volcanoes, and Tsunamis*. New York: McGraw-Hill, 2000.

Prothero, Donald R. *When Humans Nearly Vanished: The Catastrophic Explosion of the Toba Volcano*. Washington, DC: Smithsonian Books, 2018.

Reilly, Benjamin. *Disaster and Human History: Case Studies in Nature, Society, and Catastrophe*. 2nd ed. Boone, NC: McFarland, 2022.

Robichaux, Hubert R. "The Maya Hiatus and the A.D. 536 Atmospheric Event." In Joel D. Gunn (Ed.), *The Years Without Summer*. Oxford: BAR Publishing, 2016, pp. 45–53.

Sarris, Peter. "New Approaches to the 'Plague of Justinian'." *Past and Present*, no. 254 (2022), pp. 315–346.

Scarth, Alwyn. *La Catastrophe: The Eruption of Mount Pelée, the Worst Volcanic Disaster of the 20th Century.* New York: Oxford University Press, 2002.

Thordarson, Thorvaldur and Stephen Self. "Atmospheric and Environmental Effects of the 1783–1784 Laki Eruption: A Review and Reassessment." *Journal of Geophysical Research*, vol. 108, no. D1 (2001). doi:10.1029/2001JD002042.

Toohey, Matthew *et al.* "Climate Change and Societal Impacts of a Volcanic Double Event at the Dawn of the Middle Ages." *Climatic Change*, vol. 136 (2016), pp. 401–412.

Van der Brug, P. H. "Malaria in Batavia in the 18th Century." *Tropical Medicine and International Health*, vol. 2, no. 9 (1997), pp. 892–902.

Young, Bailey K. "Climate and Crisis in Sixth-Century Italy and Gaul." In Joel D. Gunn (Ed.), *The Years Without Summer.* Oxford: BAR Publishing, 2016, pp. 35–42.

Zebrowski, Ernest, Jr. *The Last Days of St. Pierre: The Volcanic Disaster That Claimed Thirty Thousand Lives.* New Brunswick, NJ: Rutgers University Press, 2002.

Works Cited

Ambrose, Stanley H. "Late Pleistocene Human Population Bottlenecks, Volcanic Winter, and Differentiation of Modern Humans." *Journal of Human Evolution*, vol. 34 (1998), pp. 623–651.

Behringer, Wolfgang. *Tambora and the Year Without a Summer.* Cambridge: Polity Press, 2019.

Bentzen, Jeanet Sinding. "Acts of God? Religiosity and Natural Disasters Across Subnational World Districts." *The Economic Journal*, vol. 129 (2019), pp. 2295–2321.

Blong, R. A. *The Time of Darkness: Local Legends and Volcanic Reality in Papua New Guinea.* Seattle, WA: University of Washington Press, 1982.

Brooke, John L. *Climate Change and the Course of Global History: A Rough Journey.* New York: Cambridge University Press, 2014.

Büntgen, Ulf *et al.* "Cooling and Societal Change During the Late Antique Ice Age from 536 to 660 A.D." *Nature Geoscience*, vol. 9 (2016), pp. 231–236.

Cashman, K. V. and G. Giordano. "Volcanoes and Human History." *Volcanology and Geothermal Research*, vol. 176 (2008), pp. 325–329.

Castro, Eddee. *Pinatubo: The Eruption of the Century.* Quezon City, the Philippines: Phoenix Publishing House, 1991.

Chester, David K. and Angus M. Duncan. "The Bible, Theodicy, and Christian Responses to Historic and Contemporary Earthquakes and Volcanic Eruptions." *Environmental Hazards*, vol. 8 (2009), pp. 304–332.

Danvers, F. C. *A Century of Famine.* Calcutta, India, 1877.

De Boer, Jelle Zeilinga, and Theodore Donald Sanders. *Volcanoes in Human History.* Princeton, NJ: Princeton University Press, 2002.

De Jong Boers, Bernice. "Mount Tambora in 1815: A Volcanic Eruption in Indonesia and its Aftermath." *Indonesia*, vol. 60 (1995), pp. 37–60.

Funiciello, R., G. Giordano, and D. De Rita. "The Albano Maar Lake (Colli Albani Volcano, Italy): Recent Volcanic Activity and Evidence of Pre-Roman

Age Catastrophic Lahar Events." *Journal of Volcanology and Geothermal Research*, vol. 123 (2003), pp. 43–61.

Gaillard, Jean-Christophe. "Was It a Cultural Disaster?: Aeta Resilience Following the 1991 Mt. Pinatubo Eruption." *Philippine Quarterly of Culture and Society*, vol. 34, no. 4 (2006), pp. 376–399.

Gerlach, Terry. "Volcanic Versus Anthropogenic Carbon Dioxide." *Eos*, vol. 92, no. 24 (2011), pp. 201–203.

Jones, Elizabeth. "Climate, Archeology, History, and the Arthurian Tradition: A Multiple-Source Study of Two Dark-Age Puzzles." In Joel D. Gunn (Ed.), *The Years Without Summer*. Oxford: BAR Publishing, 2016, pp. 25–34. Keys, David. *Catastrophe: An Investigation into the Origins of the Modern World*. New York: Random House, 1999.

Nooren, Kees *et al.* "Explosive Eruption of El Chichón Volcano (Mexico) Disrupted 6th Century Maya Civilization and Contributed to Global Cooling." *Geology*, vol. 45, no. 2 (2017), pp. 175–178.

Oman, Luke, Alan Robock, Georgiy L. Stenchikov, and Thorvaldur Thordarson. "High-Latitude Eruptions Cast Shadow Over the African Monsoon and the Flow of the Nile." *Geophysical Research Letters*, vol. 33 (2006). doi:10.1029/2006GL02766.

Peregrine, Peter N. "Climate and Social Change at the Start of the Late Antique Little Ice Age." *The Holocene*, vol. 30, no. 11 (2020), pp. 1643–1648.

Post, John D. *The Last Great Subsistence Crisis in the Western World*. Baltimore, MD: Johns Hopkins University Press, 1977.

Reilly, Benjamin. "Revisiting Bedouin Desert Adaptations: Lactase Persistence as a Factor in Arabian Peninsula History." *Journal of Arabian Studies: Arabia, the Gulf, and the Red Sea*, vol. 2, no. 2 (2012), pp. 93–107.

Scarth, Alwyn. *Vulcan's Fury: Man Against the Volcano*. London: Yale University Press, 1999.

Seitz, Stefan. "Coping Strategies of an Ethnic Minority Group: The Aeta of Mount Pinatubo." *Disasters*, vol. 22, no. 1 (1998), pp. 76–90.

Shimizu, Hiromu. *Pinatubo Aytas: Continuity and Change*. Manila, the Philippines: Ateneo de Manila University Press, 1989.

Sigl, M. *et al.* "Timing and Climate Forcing of Volcanic Eruptions for the Past 25,000 Years." *Nature*, vol. 523 (2015), pp. 543–549.

Simkin, Tom, and Richard S. Fiske. *Krakatau 1883*. Washington, DC: Smithsonian Institution Press, 1983.

Stommel, Henry, and Elizabeth Stommel. *Volcano Weather: The Story of 1816, the Year Without a Summer*. Newport, RI: Seven Seas, 1983.

Thompson, E. P. "The Moral Economy of the English Crowd in the Eighteenth Century." *Past and Present*, no. 50 (1971), pp. 76–136.

Vermij, Rienk. "Subterranean Fire: Changing Theories of the Earth During the Renaissance." *Early Science and Medicine*, vol. 3, no. 4 (1998), pp. 323–347.

Winchester, Simon. *Krakatoa: The Day the World Exploded, August 27, 1883*. New York: HarperCollins, 2003.

Witze, Alexandra, and Jeff Kanipe. *Island on Fire*. New York: Pegasus Books, 2014.

Wood, Gillen D'Arcy. *Tambora: the Eruption that Changed the World*. Princeton, NJ: Princeton University Press, 2014.

5 Climate Change

Hot, dry, and remote, the town of Talara on the western tip of South America might seem like an unlikely destination for birdwatching. But American ornithologist Robert Cushman Murphy knew that this Peruvian port was a birder's paradise, well worth the long steamer trip from Lima. Near Talara, upwelling cold-water currents brought abundant nutrients to the surface, attracting bountiful schools of fish, which in turn sustained a thriving local population of resident seabirds. Murphy also wanted to be in Talara in January of 1925 for the region's annual "Niño" event, during which the cold-water Humboldt Current off Talara was normally supplanted, briefly, by a warmer current from the north, bringing dramatic changes to local oceanic life.

Little did Murphy know that he was about to witness one of the most severe El Niños in recent memory. When he and his companions first arrived in Talara, the weather had been sunny and dry, and the cool Pacific swells were almost invisible beneath the feathers of the "unbelievable flocks of guano birds" gorging themselves on the rich local fisheries. By January 19, 1925, however, the wind began to blow in from the west, bringing with it a layer of warm, tropical water that smothered the region's colder seas. Hammerhead sharks, manta rays, and flying fish now appeared in "exceptional abundance off of Talara," and splatters of rain began to fall, an almost unheard-of circumstance along this parched coast. In the following days, clouds "dark as volcanic smoke" gathered above Talara, and the ocean temperature rose rapidly from 18°C to 26°C. Then the real rains began, downpours that turned Talara's normally hard-baked soils into a "horrible, gluey mess." Entire lakes appeared in what was normally a desert, creating breeding habitat for mosquitoes, which appeared "in such swarms as had never been known."

While Murphy was stunned by the severity of the 1925 rains, the "weather-beaten skippers and port masters" of Talara assured him that such deluges had happened before. Back in 1891, Murphy would learn, rain had poured down in "inconceivable torrents" for weeks, to the point that the "desert became a garden." Abundant vegetation had spawned

DOI: 10.4324/9781003436805-6

"swarms of insects" of "ever-changing variety and in every size and shape imaginable," and this potential prey had attracted a "great variety of spiders," including species never seen before and rarely seen again.

As a naturalist, Murphy was fascinated and delighted by the changes that the 1925 El Niño had wrought on the region's life forms and landscape. For the Peruvians, however, the events of 1925 were an unmitigated disaster. In the mountains above Talara, flash floods washed out railroad bridges and sent track and trestle tumbling into gullies. Worse yet, Talara's warmer and nutrient-poor seas proved to be fatal for the local fish, to the point that by the end of January, "dead fish, mingled with untold numbers of birds, strewed the tide lines for hundreds of miles." With their decomposition, the "ocean turned sour," blackening the bottom of local fishing boats and emptying the nets of Talara's fishermen.

But it could have been far worse. If there is any silver lining to this tragedy, it is that the climate impact of the 1925 El Niño was limited to South America. Just a generation earlier, a series of powerful El Niños had conspired with European imperialism to kill as many as 30–60 million people in three different continents, a mass mortality episode that Mike Davis (2001) has dubbed the Late Victorian Holocausts.

Between Determinism and Declensionism: Historians and Climate Change

Any study of history and climate change must begin by addressing the elephant in the room, the work of early twentieth-century scholar Ellsworth Huntington (1913). Unlike most historians of his day, Huntington argued for the importance of climate as a factor in historical developments, in two distinct ways. First, drawing on an older tradition dating back to Greek, Arabic, and European Enlightenment scholarship, Huntington argued that different world climates had left their imprint on the natural character of the world's various peoples via Darwinian evolution, creating distinct races and civilizations in the process. Second, Huntington argued that climates were not necessarily constant over the course of history, but rather could experience what he called "pulsatory changes," periods of temperature or rainfall anomalies that can be measured in decades or centuries. Huntington was, in fact, an early pioneer in tree ring analysis; not only was he one of the first historians to use tree rings to try to measure the impact of changing climate on human history, he also did much of the dirty work himself, such as counting the tree rings in 3,000-year-old California sequoias. These "pulsatory changes," Huntington argued, help to explain the rise and fall of specific cultures, since every "race" has its own ideal climate, outside of which it cannot thrive. For example, Huntington argued in 1913, "the Scandinavian does not seem to prosper greatly in the dry, sunny portions of the United States." Conversely, "the negro would apparently die out in the northern United States were he not replenished from the south."

Needless to say, Huntington's work has not aged well. Much of his science was bad, even by the standards of the time; one of Huntington's reviewers, for example, declared his book *Climatic Changes* to be "as far from being scientific as *Alice in Wonderland*." What is more, Huntington's racist and Eurocentric assumptions meant that his ideas could and would be used to justify western imperialism, eugenics, and race-based immigration laws. More broadly, Huntington was accused of environmental determinism, in other words the belief that environmental influences alone play a causative role in shaping human affairs, to the exclusion of cultural, institutional, geopolitical, technological, and other factors. Even today, scholars seeking to link human events to climate fluctuations or other natural factors run the risk of being labeled "environmental determinists" by their colleagues, a legacy of Huntington's long historiographical shadow.

In any case, by the end of World War II, the pendulum had begun to swing in the other direction, downplaying the role of climate and climate change in driving history and prioritizing instead the importance of socioeconomic, technological, and cultural factors. In an influential paper, for example, Thorkild Jacobsen and Robert Adams (1958) ascribed the progressive collapse of Mesopotamian agriculture, not to climate change, but to human choices, especially the implementation of an intensive irrigation regime that, over time, doomed the region to sedimentation and salinization. Another sign of the times was Donald Worster's (1979) monograph on the American Dust Bowl. Worster blamed this climate catastrophe, not on bad weather, but on a capitalist agricultural system that discounted the conservation of resources and instead "deliberately, self-consciously, set itself [to the task] of dominating and exploiting the land for all it was worth." Both works are classics of what has been called the "declensionist narrative" of environmental history, which seeks to explain the decline of agricultural systems, fisheries, forests, and other zones of human/environmental interaction primarily in terms of human agency. This declensionist narrative of history was further strengthened by rising public awareness of global warming in the late twentieth century. For many, the global warming crisis serves as the quintessential example of how human beings undermine the basis of their own well-being by disrupting the health of the natural systems upon which societies depend.

This is not to say that historians entirely abandoned the study of climate change during the second half of the twentieth century. Some scholars of the French *Annales* School, a group of historians with a shared interest in long-term developments in European social history, did note that changes to the climate were occurring over time, for example, by tracking the start date of the grape harvest or shifts in the location of glaciers. But even Le Roy Ladurie, the *Annales* historian most interested in climate change, dismissed the impact of climate on human history in the late 1960s as "slight, perhaps negligible, and certainly difficult to detect." For several decades, Christian Pfister (2010) points out, "this claim by the founding father of historical climatology served as a key argument to shun attempts to assess societal dimensions of climatic change." At least within history departments, therefore, the determinists were in eclipse.

Outside of the history department, however, a growing body of paleoclimatic (past climate) studies began to accumulate in the last decades of the twentieth century. Scholars in the natural sciences refined older tools of historical climate analysis, such as tree ring research, and added new techniques to their repertoire, such as the study of fossilized pollen in lake sediments, oxygen isotopes in cave stalagmites, and trapped gases within glacial ice cores. In 1982, British climatologist H. H. Lamb made an early attempt to apply some of these findings to history in his book, *Climate, History, and the Modern World*. One of Lamb's most crucial findings was that Huntington was right, at least in a limited and accidental sense: according to paleoclimate data, climate had indeed undergone a number of "pulsatory changes" over time. Lamb's data suggested that European weather had alternated between eras of warmth (such as roughly 200 BCE–200 CE and 900 CE–1300 CE, a period he dubbed the "Medieval warm epoch") and periods of colder temperatures (such as 400–900 CE and during the "Little Ice Age" of 1300–1850). What is more, unlike many of his colleagues who in the late 1970s were worried about declining global temperatures, Lamb correctly forecast that increased emissions of CO_2 during the industrial era would likely lead to significant global temperature increases by the end of the twenty-first century.

Until recently, most academic historians have been hesitant about incorporating this literature into their own studies of the human past, and understandably so. Paleoclimatology is a specialized field using its own technical jargon, so engaging with it requires historians to spend precious research hours learning at least the basics of a different discipline. What is more, most history departments in the United States have been influenced by the "cultural turn" in the field, which at its extremes has regarded science more as a bundle of discourses than as a system of inquiry, fueling skepticism about the authority of scientific claims. In addition, many historians still harbor lingering anxieties about biological and environmental "determinism," which can manifest itself as hostile exchanges at academic conferences or in inter-departmental disputes in promotion and hiring cases.

However, in the last decade and a half, the mood seems to have changed. In the early 2000s, when I first started doing research on disasters and disaster studies, few historians were paying much attention to possible links between climate change and historical events. It was a sign of the times that one of the best books on the subject, Mike Davis's (2001) *Late Victorian Holocausts*, was produced by an activist/journalist without a

PhD whose research was funded by a MacArthur Fellowship grant to independent innovative thinkers rather than through an academic posting. Since 2010, however, important scholarly monographs by respected historians, including John Brooke (2014), Sam White (2011), and Geoffrey Parker (2013), have pushed the study of climate change into the historical mainstream. To some degree, the ground was prepared for the acceptance of these studies by growing awareness about the global warming phenomenon. Most people now realize that our planet's future climate is likely to be very different from its present one, which has made people (including academic historians) correspondingly more open to the possibility that past climates have varied as well.

When writing about the climates of the past, however, historians have had to grapple not only with the legacy of Huntington but with a complex network of scientific processes. The climate, both past and present, is the result of an enormous number of variables which work on different time scales, ranging from geological eras and centuries to decades, years, and seasons. In order to make the underlying scientific processes clearer, the remainder of this chapter will be divided into three broad sections. First, we will consider *shorter-term* climate fluctuations, including the Pacific Decadal Oscillation (PDO), the El Niño/Southern Oscillation (ENSO), and the North Atlantic Oscillation (NAO), as well as scholarship relating those processes to historical phenomena. Second, we will look at processes leading to *longer-term natural* climate changes, principally through variation in solar radiation over the span of decades and centuries. Finally, this chapter will look at the work of modern-day declensionists who have studied long-term pulses of *anthropogenic* (man-made) climate change, especially but not exclusively in regard to global warming.

Shorter-Term Climate Oscillations and History

The weather we experience in any particular place and any particular time is subject to a bewildering number of inputs. Some of these factors are static and predictable, such as elevation, latitude, and geographical layout. Others are temporarily variable, but fairly predictable, such as annual seasonal variation as the Earth orbits the sun. What this section will focus on is some important variable factors that are recurrent but less predictable, including major changes or "oscillations" that unfold over the scale of years or decades and can bring significant alterations to the weather over large areas. At this point, climate scientists are fairly certain about the existence of about at least nine such oscillations, which between them span most of the globe, and which interact with each other in complex ways. They can also interact with human vulnerabilities to create disasters, as we shall see repeatedly below.

Several of these oscillations occur on the scale of decades. When the Pacific Decadal Oscillation (PDO) is in a "positive" phase, for instance, temperatures in the northwestern Pacific are considerably lower than normal and the Pacific coast of North America experiences greater heat and humidity, but the US South gets below-average rainfall. The PDO also impacts ocean temperatures, which in turn can affect the populations and distribution of commercially-important fish; indeed, the dramatic impact of the PDO on Pacific salmon populations is what brought this phenomenon to the attention of the scientific community in the first place. After staying fixed in a "positive" mode for 20 or 30 years, the PDO can abruptly switch to negative, leading to the opposite effects on weather and fisheries. Although the PDO's impacts on history are still poorly understood, Benson et al. (2007) have claimed that the multi-decade droughts which helped to trigger the collapse of the Anasazi, Fremont, Lovelock, and Mississippian Cahokian societies from the eleventh to the thirteenth centuries CE were likely exacerbated by shifts of the PDO into the negative phase, which tends to bring drought to the American interior. Benson et al. also suggest that a second multi-decadal oscillation, the Atlantic Multidecadal Oscillation (AMO), may have further worsened these droughts, since a positive-phase AMO suppresses precipitation in roughly the same areas. Nor is the modern USA immune to the same effects: a study by Jeffrey A. Lee and Thomas E. Gill in 2015 found that ocean temperature anomalies linked to the PDO and the AMO played a role in the multi-year drought that triggered the Dust Bowl. While the Dust Bowl drought was undoubtedly severe, Lee and Gill note, it was "neither unusual in the region nor extreme in length from a climatological perspective."

The AMO, in turn, is strongly linked to a shorter-term oscillation that affects the weather in the same region, the North Atlantic Oscillation (NAO). The NAO, which oscillates on the scale of years rather than decades, is measured by calculating the differential between the Greenland low pressure zone and the Azores High. When the NAO is in a positive mode, the differential between these zones is strong, which prevents the jet stream current in North America from tracking southwards. As a result, the eastern US experiences warm weather, northern Europe's weather is warm and wet, while the lands adjacent to the Mediterranean become dryer. The NAO's weak phase, in turn, has roughly the opposite climate impacts.

It should be noted that while the NAO has "positive" and "negative" modes, it should not be thought of as a light switch with only two positions, "on" or "off." Rather, the oscillations of the NAO, like those of the PDO and AMO, are better likened to a dimmer switch which can be set to fractional values along a continuum. In any given year, for example, the NAO could therefore be neutral, weakly negative, strongly positive, etc. According to Brian Fagan (2000), when the NAO is in a strong positive mode, the growing season in Scandinavia can be as much as 20 days longer than during a neutral or negative NAO phase. At the same time, a strongly positive NAO can lead to severe coastal flooding in Europe. A case in point was the Saint Marcellus's Flood of 1362, during which severe winter storms coincided with full moon high tides, creating floods that ravaged coastlines throughout Northern Europe. In coastal Holland, the death toll from these floods was so high that they are remembered as the *Grote Mandrenke*, or the "Great Drowning of Men."

If a recent paper by anthropologist B. Lee Drake (2017) is correct, the NAO may have also played an important though indirect role in the collapse of the western Roman Empire. According to Drake, all four of the largest migration episodes of Germanic peoples into Roman territory coincided with a weakening of the NAO, which would have triggered food instability in cold northern Europe through drought and a shorter growing season. Drake notes that the first two Germanic invasions did not topple Rome, though that of the Cimbri in 113–101 BCE served to militarize Roman society and might have helped transition Rome from republic to Empire. However, the third invasion, that of the Goths in 376–410 CE, was instrumental in undermining Roman power in the western Mediterranean. As we shall see below, Drake is not the only author who claims to have found the fingerprints of climate change on the fall of the Roman Empire.

The Science of the ENSO

The best-understood, and perhaps most historically significant, of the known climate oscillations is the ENSO, the El Niño/Southern Oscillation. El Niño gets its name from the annual weather phenomenon that Robert Cushman Murphy traveled to Talara to witness at the start of this chapter. During most of the year, equatorial winds peel the top layer of the sea away from the Peruvian coast and blow it westwards, allowing for the upwelling of colder, deeper, and nutrient-rich waters. However, these normally predictable "trade winds" would occasionally weaken or even reverse directions in the winter season, briefly transforming both land and sea along the American equatorial coast in the process. Locals had dubbed this event "El Niño" (meaning the Christ child) since they noted that it often coincided with Christmas.

But El Niño is far from just a local meteorological curiosity. Rather, the visitation of El Niño to the Peruvian coast is just the most visible manifestation of a much larger climatic oscillation that impacts the weather over large portions of the planet. The equatorial trade winds which slacken during an El Niño episode are actually the bottom layer of a vast atmospheric convection cell, called the Walker Circulation after its discoverer, Gilbert Walker. In normal years, these winds push sun-warmed water westward across the Pacific Ocean, which accumulates in the vicinity of the Indonesian Archipelago. As a result, the water level in the Western Pacific can be as much as half a meter higher than it is off the coast of Peru. Peru experiences cool and dry conditions due to the upwelling of deep-ocean water, but thunderhead clouds tower over Indonesia, which is simmered and steamed by the stacked-up tropical waters that surround it.

However, during an El Niño episode, these normal conditions are effectively reversed (Figure 5.1). The Walker Circulation slows, weakening the trade winds. This allows the bulge of water that had been built up in the vicinity of Indonesia to slosh back eastward, interrupting the upwelling of water off the Peruvian coast and potentially bringing heat and humidity to equatorial South America. In contrast, Indonesia suffers from unusually

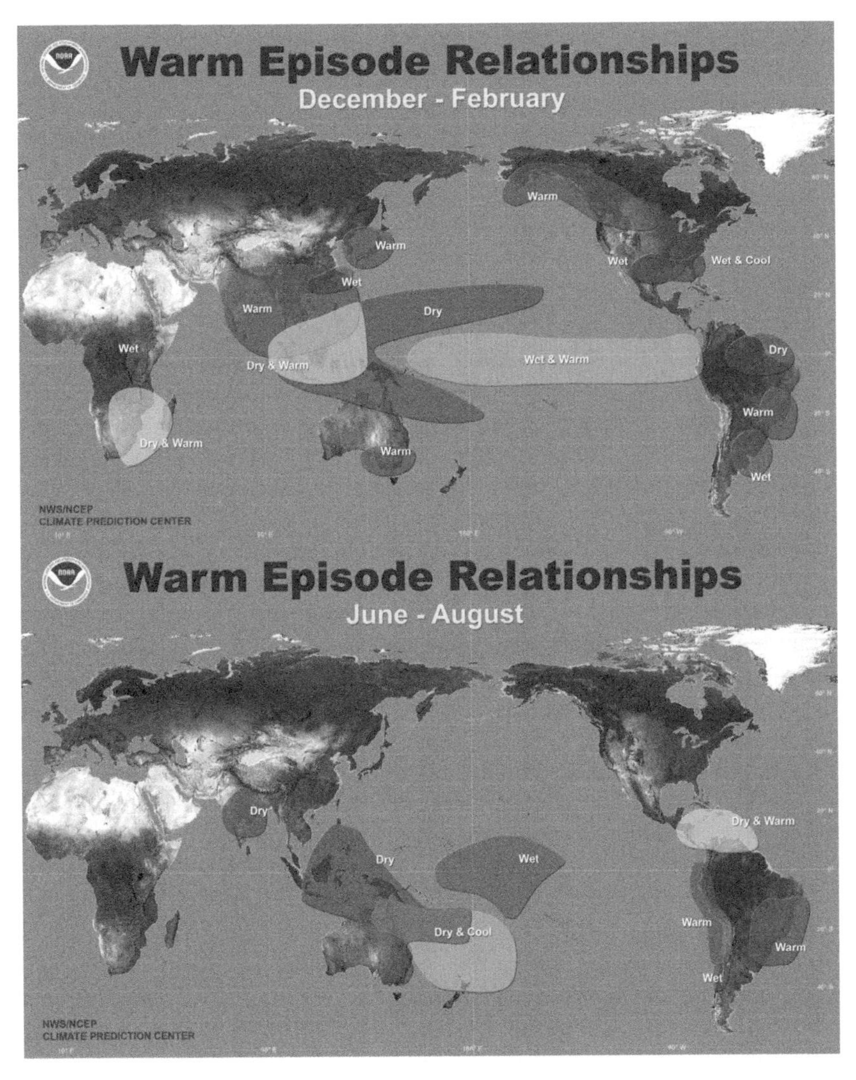

Figure 5.1 Typical weather anomalies during an El Niño event
Source: U.S. National Oceanic and Atmospheric Administration.

dry conditions and often experiences widespread wildfires during an El Niño season. Nor are the effects of El Niño limited to the equatorial Pacific Ocean. Due to teleconnections between Pacific Ocean weather systems and other climate systems, the northeastern United States and the maritime provinces of Canada experience warm winter temperatures during an El Niño season, as do coastal Alaska and western Canada, while winters in the Gulf Coast states of the southern USA are unusually cold and wet. On a more positive note, El Niño tends to reduce the number of hurricanes that strike the continental United States, though this dampening effect can be dampened in turn by high ocean temperatures in the South Atlantic. The ENSO can even impact the weather in Northern Europe, which often experiences colder winters and hotter, dryer summers during an El Niño year.

As with the other climate oscillations discussed in this section, the ENSO cycle exists along a continuum rather than a binary. If the cycle is neutral, with a Southern Oscillation Index value (SOI) calculated at close to zero, "normal" conditions prevail. During El Niño years, the SOI shifts into negative numbers, though such events can range from weak episodes, with only a mild slackening of the trade winds, to full-blown El Niño events. El Niño also has a sister, La Niña, who sits in positive values at the opposite end of the climate spectrum. During a La Niña year, conditions resemble that of a normal year, but more so; the eastern end of the Pacific becomes even cooler and dryer, and warm, moist conditions extend even farther westwards and northwards in the western Pacific.

Most importantly, the ENSO cycle impacts global weather by shifting the location of the Inter-Tropical Convergence Zone (ITCZ), potentially disrupting the annual cycle of monsoonal rains worldwide. When the ENSO is in a positive mode (La Niña), greater ocean/land temperature differentials push the ITCZ farther from the equator, usually leading to abundant monsoon rains. However, the ITCZ is pulled toward the equator during an El Niño year, weakening monsoon rainfall, and potentially bringing drought to areas that are dependent on the monsoon for a large share of their annual moisture, such as East Africa, the northeast of Brazil, parts of China, and, above all, the Indian Subcontinent. Nearly all of northern India's annual rainfall occurs during the monsoon season, and in general, the magnitude of these rains corelates closely with the Southern Oscillation Index.

That being said, there is still much about the ENSO that we do not know. While the general workings of the Southern Oscillation are pretty well understood at this point, the specific factors that trigger a given El Niño or La Niña episode are not. Forecasting El Niño effects remains frustratingly inexact as well. As Grove and Adamson (2018) point out, much of India enjoyed normal rainfall levels during the severe 1997–1998 El Niño, which was the strongest such event in a century, and farmers who had planted drought-resistant crops on the advice of meteorologists were ruined when ample rains arrived. Further complicating matters, some scientists now argue that El Niño has two modalities, its normal manifestation (as described above) and also "El Niño Modoki," from a Japanese word meaning "similar, but different." During an El Nino Modoki event, warm water from Indonesia does not migrate all the way to the eastern Pacific, but lingers near the central Pacific, and interacts with global weather through teleconnections in quite different ways.

Revolutionary Weather

One thing that we do know for sure is that when the Southern Oscillation Index is extreme, either positively or negatively, and/or it stays locked in one position for a prolonged period of time, the ENSO's weather can shape human history. Richard Grove (2007), for example, has argued compellingly that the "Great El Niño" of 1788–1795 may have played a role in the outbreak of the French Revolution. By all accounts, the weather in France went haywire in 1788–1789. A cold spring gave way to a hot, dry summer that withered the wheat on its stalk, and when rains did fall, they were accompanied by hailstorms, which "cut down wheat and trees in [their] path." As a result of the heat and hail, some areas of the French countryside were reportedly reduced to an "arid desert," and the autumn harvest was so poor that the price of bread doubled between 1787 and 1788. France's summer of troubles was followed by the coldest winter it had faced since 1709, leading to a spike in mortality. What is more, grain shortages provoked bread riots and popular assaults on warehouses and grain transport convoys, as local peasants enforced the "moral economy of the poor," much as they would a generation later during the post-Tambora volcanic winter. Hungry peasants also turned to widespread poaching; "rabbits, deer and other game were slaughtered, irrespective of ownership or regulation," Grove tells us, and "gamekeepers and other symbols of authority who opposed such actions were killed." When the revolutionary year of 1789 came around, therefore, France's rural population was already in a restive state.

The woeful weather of 1788–1789 may also help to explain the outbreak of the "Great Fear" in late July–early August 1789. During these panic-filled days, rumors spread through rural communities about sinister conspiracies against the peasants. A "famine plot" was being hatched, the peasants told each other – aristocrats were hoarding grains and had even hired foreign soldiers (English? Piedmontese?) or domestic brigands and vagabonds to destroy the crops in the field. By late July, alarms were being raised all over rural France that armies of foreign thugs in league with the aristocrats had been spotted on the march within France, plundering and murdering as they went. As a result, peasants throughout France armed themselves and turned against the rural nobility, who were threatened, beaten, or humiliated, and, in a few cases, killed, while manor houses and feudal charters were torched.

So why were wild rumors of aristocratic plots and foreign blackguards so credible to French peasants in the summer of 1789? Political drama in Paris probably played a role. In early 1789, French King Louis XVI had to been forced to summon the Estates-General, a medieval representative body, to try to solve to the French state's deep financial problems and resulting political weakness. The Estates-General consisted of three separate estates, the "First Estate" of clerical delegates, the "Second Estate" of nobles, and the "Third Estate" of the commoners, which was numerically the largest and represented the most Frenchmen but was outvoted by the other two estates. Unwilling to accept this subordinate status, the Third Estate declared itself to be the National Assembly in June, and many clergymen from the First Estate joined it. The result was a stand-off between the National Assembly and the French monarchy, and in the month before the Great Fear, Georges Lefebvre (1973) tells us, rumors were being spread that the French King and high aristocrats were gathering foreign mercenaries and domestic "brigands" to storm Paris and dissolve the Estates-General.

In Paris, these rumors helped to inspire the celebrated attack by the French revolutionaries on the Bastille fortress and prison on July 14. This show of popular strength probably reduced the internal threat that the Third Estate faced from the monarchy and aristocracy. However, in the minds of many rural French, it may have increased popular anxieties about *external* threats to the revolution. Following the destruction of the Bastille, a wave of high aristocrats fled France for oversees political exile, where these *émigrés* were said to be hiring an army of foreign mercenaries and conspiring to destroy the peasants. According to French historian Georges Lefebvre, many in France feared that "these princes cannot allow themselves to be exiled from a kingdom which is their native land and their heritage without plotting a revenge for which they would gladly sacrifice everything." As a result, it was believed, the *émigré* aristocrats were conspiring to send "foreign troops into France, of intriguing with the nobility to wipe out Paris and everything pertaining to the Estates-General." The French aristocracy, therefore, was increasingly associated in the minds of many Frenchmen with menacing foreign threats.

These rumors of an aristocratic pact with foreign brigands may also have seemed credible because the countryside was full of strangers in 1789 as a result of the poor 1788 harvest. As Lefebvre noted, in the aftermath of 1788, many "poor and needy left their own villages and wandered off for miles around," including farm workers and craftsmen who could no longer afford bread at the inflated 1788 prices. Even in normal years, beggars were a cause for anxiety or concern, since it was feared that they would raid the fields before harvest or damage farm infrastructure if their demands for hand-outs were refused. But this was not a normal year, and by the summer of 1789 many farmers were themselves starving, as early summer is traditionally the time of the *soudure*, when last season's harvest surplus was nearing exhaustion and the current year's harvest was still unripe in the fields. In addition, many rural Frenchmen were inclined to blame their hunger, not on poor harvests, but on a fiendish aristocratic plot. In eighteenth-century France, many Frenchman ascribed to the *pacte de famine* conspiracy theory, believing that famines were not natural occurrences but human inventions, the result of grain being adulterated or else deliberately withheld from the markets for the benefit of the elites. These fears may have arisen in part to the French state's policy since the 1760s of progressively relaxing earlier restrictions on the grain trade in favor of free market economics.

Rural France had ample reasons for fear in the summer of 1789, therefore, but still, the sheer scale of the Great Fear still seems to demand an explanation, especially since the "foreign brigands" that inspired widespread panic were purely a figment of the popular imagination. Mary Kilbourne Matossian (1989) has gone so far as to blame the Great Fear on mass hallucination, triggered by ergot poisoning, since the low 1788 harvest and high market grain prices obliged French peasants to eat moldy grain. Well, possibly. But ergotism does not explain why peasants throughout France were motivated by a specific set of fears, namely foreign brigands and their supposed aristocratic allies.

A simpler explanation for the same phenomenon arises from our basic human psychology, namely our predisposition to xenophobia and outgroup exclusionary behavior when the behavioral immune system is activated. While France seems to have escaped mass mortality after the 1788 harvest failures, it had not escaped widespread hunger, which, in theory, should have made people more sensitive to disease cues. Remember, according to Curtis et al. (2011), malnutrition is theorized to put the BIS into overdrive by "upregulat[ing] disgust," "potential sources of pathogens," such as food impurities, poor hygiene, bodily fluids and fecal matter. And as we have seen, the French countryside in early 1789 was plagued by gaunt, hungry, and filthy beggars who may have served as walking disease cues, helping to provoke animus against outsiders. Further evidence that the Great Fear was rooted in the 1788 famine comes from a recent study by Maria Waldinger in 2021, who has argued that towns and provinces who suffered badly from the 1788 drought were also statistically more likely to participate in peasant revolts in the summer of 1789.

Of course, this is not to say that the rioting during the Great Fear was in any way inevitable, or that the political, economic, social, and cultural factors played no role. Far from it: the political chronology I have outlined above helps to explain why peasants may have focused their general outgroup animus against a specific target, the aristocrats, who due to recent events had become associated in the popular mind with dangerous foreign brigands. What is more, French peasants had plenty of pre-existing social grievances against the land-holding aristocracy, above and beyond their rumored collusion with foreign thugs. Still, the behavioral immune system might help to explain why hungry French rural populations were so willing to believe wild rumors about non-existent armies of blackguards and assassins in the summer of 1789.

More broadly, the 1788 "Great El Niño" drought might have contributed to the French Revolution by crippling the French industrial economy and throwing thousands of urban workers into unemployment. As Lefebvre (1973) points out, even before the anomalous weather of 1788, the French industrial sector had been struggling. The recent outbreak of war between Ottoman Turkey and an Austro-Russian coalition had rendered the Baltic Sea and Eastern Mediterranean dangerous for shipping, reducing French exports. What is more, a 1786 trade agreement between France and English had reduced French tariff duties on English manufactured products, causing terrible distress in French industrial centers, which could not compete with cheap English textiles and other finished goods. The Great El Niño further exacerbated these existing problems. Industrial production in mills dependent on waterpower ground to a halt during the terrible winter of 1788–1789 when most French rivers froze solid. What is more, the precipitous rise of grain prices meant that most French family income was going toward foodstuffs by 1789, drastically reducing demand for manufactured goods. As a result, many French workers lost their livelihoods just when grain prices were spiraling upwards. Small wonder, then, that the popular crowds in Paris and other French cities were already in an agitated state when Louis XVI summoned the Estates-General.

The notion that the Great El Niño of 1788–1795 may have played a role in triggering the French Revolution through its economic impacts seems especially plausible in light of a study by Soloman M. Hsiang et al. in 2011, who claim that ENSO episodes can be statistically linked to outbreaks of civil violence in the second half of the twentieth century. In the tropics, Hsiang et al. claim, new civil conflicts are twice as likely to break out during an El Niño year compared to La Niña years, which serve in their study as a control set. Overall, they claim the ENSO may have played a role in triggering 21% of all civil conflicts since 1950, and they speculate that the ENSO exerts this impact through its adverse economic effects, "such as loss of income or increasing food prices." They also found that El Niño episodes were particularly likely to trigger civil conflict in low-income countries, though in this case they don't know whether sensitivity to the ENSO is the result of low incomes, or if low incomes are a long-term result of repeated exposure over historical time to detrimental ENSO climate effects. Either way, Hsiang et al. are right, the ENSO cycle may be intimately connected with global cycles of violence.

The Late Victorian Holocausts in British India

In addition to helping provoke revolution in France, Grove (2007) claims, the Great El Niño of 1788–1795 also contributed to terrible famines throughout the subtropical world. In Egypt, which had already been struck hard by crop failures during Laki's volcanic winter of 1783–1784 (see Chapter 4), the Nile floods were low once again in 1791 and 1792, triggering widespread famine. Egyptian sources from this period claim that people resorted to eating dead horses, donkeys, and even children. In the meantime, a similar drama was playing out in India, which had barely recovered from the severe Chalisa famine of 1783–1784 that, as we saw in Chapter 4, may have also been triggered by the sulfates of the Laki volcano. Rainfall in some parts of India plummeted from nearly 200 centimeters in 1788 to only 109.4 cm in 1789 and then to 44.2 cm in 1790, and they would remain at a low level until 1793. As many as 11 million may have died in India during this multi-year drought, and in some parts of India, the records of the British East India Company suggest that two-thirds of the villages were abandoned.

Powerful El Niño events would once again rock the subtropics at the end of the nineteenth century, and when they did, they afflicted societies that were even more vulnerable than before due to socioeconomic changes imposed upon these societies by European imperialism. Since monsoon failures were neither uncommon or nor unexpected in India, ruling regimes in pre-British India had long employed multiple strategies to prevent hunger when the monsoon did not arrive, including free distribution of raw grains and cooked foods, remission of taxes, providing employment through the building of public works, and cash payments to the afflicted population. Indeed, as Jean Drèze (1990) points out, the body of Indian literature dedicated to governmental anti-famine interventions dates back fully 2,000 years. This is not to say that India's traditional anti-famine interventions were uniformly successful during the era before the British conquest of India. As Grove and Adamson (2018) note, during the 1594–1598 famine in Northern India:

> The king [of Hindustan] ordered that alms should be distributed in all the cities and Nawab Shaikh Farid Bokhari being ordered to superintend and control their distribution, did all in his power to relieve the general distress of the people. [Despite this], in consequence of the dearth of grain and the necessities of ravenous hunger, men ate their own kind. The streets were blocked with dead bodies and no assistance could be rendered for their removal.

Indigenous regimes in India, therefore, had an imperfect record when it came to combatting hunger. But at least they did not make the famine problem worse, which cannot be said for Britain, whose control over India coincided with a series of horrific El Niño-induced famines. By 1876, the starting year of the first such famine, the British enjoyed undisputed dominion over the "Raj," which included most of the modern nations of Pakistan, India, Bangladesh, and Myanmar (Burma). Over time, the

British had re-made the economy of the Raj to fit the needs of the industrial metropolis, for example, by encouraging the cultivation of cotton, jute, indigo, and other industrial raw materials at the expense of local subsistence foods. Insofar as the British did encourage the cultivation of foodstuffs, it was for their own benefit: Britain regularly imported thousands of tons of Indian rice and grain annually, with the goal of keeping British food prices low and thus lowering the cost of labor in Britain. To get these raw materials and food to the home market, the British had started the construction of a railway network in India, which by the 1870s linked the ports of Bombay, Madras, and Calcutta to the Indian interior and were subsidized by high taxes on Britian's Indian subjects. Indian tax revenue was also used to fund a series of British military campaigns into Afghanistan, as part of Britain's "Great Game" competition with Russia over the mastery of Central Asia. None of these expenditures brought any real benefit to the average Indian subject of Queen Victoria, who suffered from stagnating or declining real wages under British rule.

As a result of self-interested British manipulation of the economy of the Raj, India was already in a state of partial crisis even before El Niño suppressed the rains in 1876. That year's monsoon was notable mostly by its absence; virtually no rain fell in much of India from June to August, the normal rainy season. As a result, crop losses in some areas ranged from 66% to 95%, and since the agricultural surplus of the previous few years had been syphoned off by overseas markets, famine set in almost immediately. According to Eugene Linden (2006), starving Indians "ate shoe leather, then livestock, then dogs, then plough animals," the consumption of which perpetuated the famine problem by putting future harvests at risk. When these food sources were exhausted, "people on occasion began to eat each other, and when they became too weak to move, dogs and wild animals ate them."

India's 1876–1878 famine did not limit itself to the worst-affected areas in India, but became general, since the railroad had effectively unified the Indian grain market. Grain prices spiked throughout India, spreading hunger even to provinces that had received adequate rainfall, and the high expense of food meant that demand for handicrafts and services plummeted, throwing non-agricultural laborers in India out of work. What is more, market forces tended to pull grain out of the provinces that needed it most, since the impoverished areas worst impacted by the monsoon failure were outbid on the market by less-affected provinces. As a result of these factors, Linden points out, India's poor "starved even more rapidly than in pre-railway droughts."

The British response to this growing famine served only to exacerbate it. Intervening in the grain markets was out of the question, since it would have been an affront to Smithian free market principles. In addition to being acolytes of Adam Smith, the British were also students of his fellow economist Richard Malthus, who believed that demographic crashes were inevitable in agricultural societies when exponential population growth outstripped the linear growth of food supplies. As a result, British officials discontinued the previous Indian government policies of distributing food and/or money in times of famine, thinking it would only perpetuate the overpopulation problem – and in any case, such hand-outs would ruin the moral character of the Indian laboring masses and create dependency on the government. But all of these objections to governmental interventions into the famine may have just been rationalizations, justifications for tight-fisted policies that prioritized lowering expenditure rather than saving lives. The whole point of controlling India, after all, was to make money, not spend it; as Mike Davis (2001) quipped, British officials regarded India "as a revenue plantation, not an almshouse."

Ultimately, the only form of famine relief the British were willing to offer was via public works projects (Figure 5.2). Starving Indians were therefore set to work laying railroad track and digging canals, for which they received a meager daily food ration that was half the recommended intake for adults and which Davis claims provided less calories than the "diet inside the infamous Buchenwald concentration camp [of World War II]." Not surprisingly, mortality rates in these work relief camps were atrocious; one British official reported to his superiors that the road relief project he visited "bore the appearance of a battlefield, its sides being strewn with the dead [and] the dying." Some impoverished Indians even begged to be arrested, since food rations in the jails were far higher than those given to workers performing hard labor in the "relief" camps.

THE FAMINE IN INDIA—DISTRIBUTION OF RELIEF TO THE SUFFERERS AT BELLARY, MADRAS PRESIDENCY

Figure 5.2 British distribution of famine relief in India, 1877
Source: *Illustrated London News*, May 26, 1877.

Conditions at the camps only improved after hundreds of thousands of camp workers went on strike and their cause was taken up by Indian middle-class nationalists, who embarrassed the British government by publicly claiming the camps were saving no lives but rather were serving only to doom tens of thousands to the "slow torture of starvation." Overall, the 1877 El Niño killed approximately 5.5–7 million subjects of the Raj.

In the aftermath of the 1877–1878 famine, British officials created a new famine code and established a Famine Relief and Insurance Fund, designed to prevent mortality on the scale of 1877–1878 from happening again. One economist of the time, Augustin Filon, predicted that these measures, combined with the further expansion of India's railway network, meant that "famine in the original sense of the word, that is to say as a result of a lack of food, has become impossible" in India. But, in 1896, the return of El Niño would prove Filon wrong. That year, monsoons failed once again in much of the Raj, bringing hunger to a population which was already malnourished after several years of relatively poor harvests. Worse yet, there was little reserve grain in India, as much of it had been exported to Britain to compensate for serious shortfalls in England's 1895 harvest. As hunger set in, the famine codes were widely ignored, and the relief fund was found to be greatly depleted, as (according to Davis, 2001) "a substantial portion" of it "had been diverted against the protests of Indians to pay for yet another vicious Afghan war."

As for the Indian railroad system, which British believed would be a bulwark against famine, it once again helped turn local harvest failures into a general catastrophe. Just as in the previous famine, the railroad network served to concentrate grain, not in areas of the greatest need, but rather the areas with the greatest purchasing power, which also tended to have the least hunger. Merchants hoarded desperately-needed grain stocks, anticipating prices would climb still higher, which of course they did in part because grain was being kept from the market. Local officials were powerless to stop this profiteering, since they had been forbidden by the government of the Raj from interfering in the sacred workings of the free market. As a result, trains full of grains passed almost daily through starving villages, and journalist Jefferson Ellsworth Scott reported that "at many of the railway stations I saw thousands of fat pigeons gorging themselves with grain from the loaded wagons at the siding … in the sight of scores of miserable, famine-stricken villagers crying aloud for food" (Davis, 2001). Worst of all, while the railroads did little to shift grain to starving areas, they did assist in the spread of epidemic diseases, particularly the gastrointestinal infection cholera.

Although the British responded to the famine by setting up poor houses in some areas, the conditions in these charitable institutions were atrocious, in part by deliberate design. As Joanna Simonow (2023) points out, British officials believed that Indians were "shirkers by nature," so to discourage them from resorting to poor houses (and thus burdening state finances), relief-seekers were obliged to perform grueling labor to earn their keep. In return, they received "poor and monotonous meals," which once again were below the minimal daily caloric requirements to maintain basic metabolic functions, much less perform hard labor. Vaughan Nash, a journalist who visited these poor houses, was shocked to find Indians working nine hours a day on little more than 400 grams of food, leading to an "enormous death-rate" at the very camps that had been established, in theory, to keep people alive. Worst of all, government expenditures on famine relief actually declined over time, as the famine spilled over into multiple years, since the British government's attention was diverted after 1899 by the Second Boer War in South Africa. Scholarly estimates for the overall death toll in India for the 1896–1902 famine in India range from a low of 3–4.4 million to *The Lancet*'s calculation of 19 million deaths, which they considered to be a conservative estimate.

More generally, Mike Davis (2001) has argued that El Niño, combined with British imperialism, may have played a role in locking the Indian Subcontinent in persistent poverty. Prior to the nineteenth century, Davis notes, Indian laborers actually enjoyed higher wages and a better standard of living than their British counterparts. The usual explanation for Britain's spike in net wealth after 1800 is the advent of the Industrial Revolution, but Davis finds this explanation to be facile, and notes that at least until the 1850s, Asia "surrendered ground very grudgingly" to British and other European competition. What secured British predominance, Davis contends, was not industrialization per se, but the early Industrial Revolution's impact on British military capacities, which put Britain in the position to dictate terms to much of Asia after 1850 and to re-make its economic systems in Britain's interest. Once in a position of dominance, Britain not only actively de-industrialized its Asian competitors, but also extracted enormous revenues from its colonies, helping to fuel its so-called "second industrial revolution" of the late nineteenth and early twentieth centuries. In the meantime, Britain's imperialist intervention in the Raj's economy undermined the ability of the subcontinent to protect itself from El Niño and famine. Staggering under the dual blows of drought and imperialist exploitation, India's share of the world GDP shrank precipitously from 1700–1950, and even today India continues to lag behind industrialized North America and Europe.

The fact that British colonial policies, and not the climate, bear primary responsibility for the "Late Victorian Holocaust" famines in Britain is amply demonstrated by India's success in combatting famine since the British departure in 1947. While India has certainly suffered from drought and poor harvests since the middle of the nineteenth century, India has not experienced a major famine since independence. Even the serious Maharashtra drought of 1972 led to no deaths from starvation. As Alex de Waal (1997) points out, post-independence India now enjoys a crucial form of protection that was lacking during the British era: an enforceable anti-famine contract between citizens and the state. Indians now feel that freedom from famine is a basic right, which is enshrined in their Constitution and defended through a free press, popular elections, and occasional direct political action like public demonstrations. In the end, India was rescued from El Niño-induced food shortages, not by railroads, free markets, or famine defense funds, but by democracy.

European Imperialism and El Niño in Qing China

As Davis (2001) points out, India was far from alone in its misery during the 1876–1878 and 1897–1900 El Niño events. Famine also stalked the monsoon-dependent Nordeste province of Brazil, the highlands of Ethiopia, and portions of North Africa during one or both periods. However, the only region where the tragic impact of El Niño may have exceeded the horrors of India was China, which may have suffered a many as 30 million deaths during the "Late Victorian Holocausts."

On the face of it, this fact seems to contradict the connection Davis draws between European imperialism and vulnerability to El Niño, since China was not a colony of Europe. However, by the mid-nineteenth century, China's independence had become extremely precarious. The Chinese Empire, ruled by the Qing dynasty of Manchu origins, entered the nineteenth century as the world's most populous and economically most prosperous state, but its military technology had begun to lag dangerously behind that of Europe, perhaps in part because Qing China had such a powerful advantage over its regional rivals and that there was little incentive for military improvements. This weakness was laid bare during the First Opium War of 1839–1842, during which a British/Indian expeditionary force inflicted humiliating defeats on the numerically superior Chinese and seized the strategically vital city of Zhenjiang, located where the Grand Canal (which linked northern and southern China) met the Yangtze River. Their commerce strangled by Britain's chokehold on China's most crucial artery of trade, the Qing were compelled to agree to the Treaty of Nanking, which forced them to cede Hong Kong to British control, open more Chinese ports to foreign trade, allow British consuls to represent the interests of British merchants in those cities, and pay a substantial indemnity to Great Britain. This show of weakness by China provoked the United States and France to demand their own concessions from China over the course of the next two years, further undermining the prestige of the Qing government.

Qing China's failures in the First Opium War also contributed to the outbreak of the Taiping Rebellion in 1850. Defeat at the hands of western "barbarians" had discredited the Manchus, and the Qing were no longer able to prevent shipments of addictive opium from flooding into China, creating a serious trade imbalance between China and the western powers that destabilized the Chinese economy. As a result, many southern Chinese were attracted by the ideology of the Taiping "Heavenly Kingdom" movement, which combined anti-Manchu, nationalist, and reformist elements with a syncretic religious system that blended elements of Christianity, Daoism, Confucianism, and messianic doomsday fervor. The civil war fought between the Qing and the Taiping revolutionaries would drag on until 1864, and it was a sign of the times that the Qing relied on the assistance of the superior armed forces of France and Britain to finally subdue the Taiping. Ultimately, the Taiping Rebellion resulted in 20 to 30 million deaths and severely weakened the Qing dynasty. Worst yet, in the midst of the conflict with the Taiping, Qing armies once again suffered humiliating defeat at the hands of invading European punitive expeditions, which during the Second Opium War of 1856–1860 seized the Qing capital of Beijing and burned down the Emperor's Summer Palace. The Qing were therefore forced to agree to yet more concessions to the British and French, including a massive indemnity that contributed to the Qing state's growing fiscal crisis.

The Qing state, therefore, was already tottering when the 1876–1878 El Niño weakened the normal summer monsoon in China, which that year penetrated China only as far as the Yangtze River and left the north high and dry. As was the case in India, monsoon failures of this sort were an expected occurrence in China, and, due to long experience, the state had established a set of institutions and administrative practices to mitigate the loss of life when they occurred. A serious El Niño-linked monsoon failure in 1734–1735, for example, had caused considerable suffering, but relatively few deaths, as the Qing state acted decisively to alleviate the famine. During that crisis, the Qing provided food relief from local "ever-normal granaries," where the government routinely collected surpluses from years of plenty for distribution during times of dearth. When those stocks proved unequal to the crisis, the Qing brought in enormous stores of rice from southern China via the Grand Canal, ultimately providing enough food to sustain two million peasants until the next year's harvest. As Davis points out, at a time when most European states had only rudimentary tools at their disposal to fight harvest failures, the Qing state in its eighteenth-century heyday considered famine relief to be a basic principle of good government and took proactive steps to prevent widespread starvation.

When the monsoon failed in 1876, however, Qing China was nowhere near its heyday. By the late 1870s, the Qing were deeply in debt to European lenders and had just emerged from a brutal and expensive war in Xinjiang, on the western edge of the Chinese state. The Qing's limited gains in Xinjiang, however, were counterbalanced by losses to the south, where France was carving out a colonial state in Indochina, an area traditionally under Chinese suzerainty, and in the north, where the Qing had to cede large expanses of territory to the expanding Russian state.

Arguably, however, the greatest threat to China in this period ran through the very center of the country. In 1855, during the chaotic days of the Taiping Rebellion, the Qing state had lost control over the Yellow River, which had broken free from the restraining levees and dikes. For 15 years, the sediment-laden Yellow River wandered over a large area of the Northern Chinese Plain, changing courses constantly as it choked itself on its own sediment, inflicting enormous damage to life and property in the process. It would take the Chinese until the early 1870s to finally regain control over the Yellow River, and due to the state's financial constraints, the best they could do was to confine the Yellow River in its new course to the Bo Hai Sea. As a result, the Grand Canal, China's vital north-south economic lifeline, was rendered inoperable, greatly complicating Qing attempts to alleviate the 1877–1878 famine.

The crisis first manifested in the Shandong Peninsula, which had already suffered three relatively lean years in a row when El Niño struck. Missionaries reported that starving peasants were resorting to eating "grain-husks, potato stalks, and elm bark, buckwheat stalks, turnip leaves, and grass seeds." These dubious sources of sustenance were soon exhausted, so desperate peasants sold their clothes and children, ripped out and sold the timber of their houses, and even tried to eat the sorghum stalk thatch of their roofs. When the cold season came, starving peasants were forced to burn what was left of their houses for warmth: as Davis recounts, the American missionary Samuel Wells Williams long remembered the pitiful sight of Shandong peasants "hovering over the ashes of their burnt houses, and making pyres for themselves out of the ruins of their temples." Other freezing and starving peasants resorted to the "extraordinary stratagem" of living *en masse* in giant underground pits, where they huddled together for warmth.

While Shandong's misery was particularly well documented, as it was a hotspot for western missionary activity, Shandong was far from alone in its suffering. Famine also gripped Shanxi, Hubei, Henan, Anhui, and Jiangsu, an area that Davis notes was "larger than France" and hosted about 90 million people. In Shanxi, the provincial governor reported to Beijing that the progression of the famine could be measured by its escalating scenes of horror: at first "the living fed upon the bodies of the dead," then the "strong devoured the weak," and "now, the general destitution has arrived at such a climax that men devour those of their own flesh and blood." This grim report was confirmed by a contemporary missionary, who wrote of "men and women lying helpless on the roadside," the "dead torn by hungry dogs and magpies," and dark rumors of children being "boiled and eaten up."

Chinese people in the famine-afflicted regions responded to the crisis in various ways, for example, by relying on collective "blood-oath" fraternities for support, putting pressure on local officials for distribution of food relief, joining religious cults and rebellions, and engaging in "t'aofang" traditions of organized migration and collective begging. Given the scale of the shortages, however, only large-scale state relief could have averted mass casualties. But the Qing state of 1876–1878 was not the Qing state of 1734–1735. Thanks to western indemnity payments, the economic damage inflicted by the Taiping Rebellion, the cost of the Xingjiang war, and the ruinous interest payments on their foreign debts, the Qing government was in no position to act decisively at this time of crisis. The situation of the provincial governments was even worse: many local governments in China were already bankrupt even before the start of the drought, and poverty or corruption had emptied China's "ever-normal granaries." While the Qing state did try to provide food relief to its hungry northern provinces, its ability to do so was hampered by both financial constraints and serious transportation problems, especially the closure of the Grand Canal. In the end, many inhabitants of the drought-stricken provinces did not wait for food to come to them, but set out to find it, migrating in the hundreds of thousands to major cities. But assuming they survived the hungry, harrowing journey, many refugees discovered they had only flopped from the frying-pan into the fire, as diseases of poverty like typhus and cholera killed hundreds of malnourished refugees a day in the squatter camps of Beijing and Tianjin. In all, the death toll from the Northern Chinese Famine of 1876–1879 has been estimated at between 9.5 and 13 million people, including over 5 million in Shanxi alone, where one out of three inhabitants may have perished.

El Niño, the Boxers, and Xenophobia during the Late Qing

While the 1876 El Niño and subsequent famine were a tragedy for the Chinese, many Christian missionaries operating in China saw it as an opportunity for evangelization. The Convention of Beijing after the Second Opium War had given Christians the right to do missionary work in China, but the Chinese had proven to be stubbornly resistant to conversion. When the 1877 famine began, therefore, many missionaries changed tack and turned to massive food relief as a means of selling salvation to the Chinese masses, in the belief that the best way into a Chinese man's soul was through his stomach. This same materialistic approach to proselytization continued even after the famine had abated, when missionaries provided food, medical care, and financial benefits to Chinese Christian converts. Missionary groups also supported Chinese Christians in legal disputes against non-Christian converts by appealing to the powerful western consuls in major port cities. This fueled a growing resentment against Christian Chinese, who were increasingly seen as collaborators with predatory foreign powers. Be that as it may, by the turn of the century, missionaries were beginning to enjoy some success in their project of evangelization. In 1900, according to Diana Preston (2019), China hosted about 700,000 Catholic and 85,000 Protestant Christian converts.

Christian missionary success, in turn, helped to spur the creation of a new quasi-religious society in Shandong, the Righteous and Harmonious Fists. Preston argues that the Righteous and Harmonious Fists, like the Taiping rebels of a generation before, followed syncretic religious beliefs that "fused religion, theatre, magic, and martial arts." This public martial arts training, in fact, is what inspired Shandong's many missionaries to call this group the "Boxers," the name by which they are commonly known today. According to Preston, while the Boxers sought spiritual solutions to China's problems, the roots of the movement were largely economic. By and large, the Boxers were recruited from the "poor and dispossessed of Northern China," whose lives had "long been a losing struggle against cycles of flood, drought, and famine." What is more, the closure of the Grand Canal had struck Shandong hard. Before becoming inoperable, the canal had helped to stave off flooding in Shandong and lowered food prices by bringing in rice from Southern China, benefits that were now lost. In addition, many people in the province had previously earned their livelihood as boatmen or maintenance workers before the canal closed.

Broadly speaking, the Boxers consisted of two affiliated groups. The "Spirit Boxers" were the spiritual wing of the Boxers, faith healers, and shamans who claimed that their spirit possession rituals rendered them invulnerable from harm. The "Big Swords," on the other hand, were in essence a peasant militia group organized to combat the bandit groups that were proliferating in China due to the weakness of the Qing state. Both groups also opposed western missionaries, especially after 1895, when a group of bandits captured by the Big Swords claimed to be Christians in a ploy to escape prosecution, an increasingly common tactic of the time that the Chinese referred to resentfully as "relying on foreign strength." As result, in the words of historian Paul Cohen (1997), "the line between Christians and bandits became increasingly indistinct."

These growing tensions would explode into xenophobic violence when ENSO-related climate instability returned to China at the turn of the twentieth century. In 1896, the monsoons again failed to reach northern China. While the hunger it caused was not as severe as the famine that was gripping India during the same year, Davis (2001) argues that anxieties triggered by the drought became "the bellows that transformed local sparks of antiforeignism into a vast populist conflagration across north China." One such spark was of course the Boxers of Shandong, who explicitly linked famine with the work of Christian missionaries. "No rain comes from Heaven," Boxer propaganda posters proclaimed, and "the earth is parched and dry. And all because the churches have bottled up the sky."

Popular anger against Christian missionaries culminated in the Juye Incident on November 1, 1897, just weeks after Shandong's peasants had gathered in the meager harvest that had survived the El Niño monsoon failures. On that night, Chinese assassins attacked a Christian mission in Shandong, vowing to kill the missionary priest George Maria Stenz, who was rumored to have raped several Chinese women. While they failed to kill Stenz, they did murder two other German Catholic missionaries. In response, Germany flexed its military muscle and forced China to hand over control of large portion of Shandong, where the Germans began construction of a naval base. In the process, the Germans further outraged Chinese sensibilities by constructing railroads and telegraph lines atop of Chinese burial sites, and when popular protests erupted, they were savagely repressed by the Germans. Chinese anger was further fueled when 1897 turned to 1898 and drought gave way to devastating floods, which many Chinese blamed on the corruption of local Qing officials. The head of the local flood control board was dismissed by the central government for his failures, but was restored to power after proclaiming his pro-Catholic sympathies to the French. As a result, Chinese public opinion now blamed foreign interference as much as Qing incompetence for the destructive 1898 floods.

China was now primed for an explosion, and when the destructive floods of the 1898 season gave way to returned monsoon failure in 1899, the effect (in the words of Davis, 2001) was "like throwing a match into a pool of gasoline." Chinese fear of another famine on the scale of 1877–1879 strengthened and further radicalized the Boxers, who began to preach the slogan "equal division of grain" and turned their sights on groups that had more than their fair share of provisions, including merchants, rich peasants, and the well-supplied "rice Christians" at the missions. As was the case in 1897, the return of drought was blamed on foreigners, and it was predicted the drought would not end until Christian influences were rooted out in China. What is more, wild anti-foreign rumors began to spread in northern China. Not only were missionaries poisoning the wells of Chinese villages, it was claimed, they were collecting and drinking Chinese menstrual blood and kidnapping Chinese children to steal their hearts. Some asserted that foreign vessels were daily departing from China with holds filled with Chinese blood, women's nipples, and even human eyes, which were said to be needed by westerners for their devilish art of photography.

Radicalized by famine fears and wild rumors, the Boxer movement soon rose to an anti-Western crescendo. In late 1898, the Boxers began to preach a new slogan, "support the Qing, destroy the foreigners". At first, the Qing tried to suppress the Boxers, but when support for the Boxers swelled in 1899, the Qing changed its mind and began to issue pro-Boxer edicts. Encouraged by official sanction, the Boxers began to attack European Christians (whom they referred to as the "great hairy ones") and Chinese converts (the "secondary hairy ones"), first in Shandong and then in the vicinity of Beijing. Both groups were subjected to atrocious acts of violence. Contemporary accounts describe "women and children hacked to pieces, men trussed like fowls, with noses and ears cut off and eyes gouged out." Chinese Christians were reportedly "skinned alive, set alight, or buried still living," and in one case, a missionary was brutally executed after being found guilty by a Boxer tribunal for a laundry list of crimes, including "poisoning the wells, causing the drought, even using [Chinese] babies as fish bait." Ultimately, more than 200 Catholic and Protestant missionaries in China were killed by the Boxers, along with their wives and children, and over 30,000 Chinese converts to Christianity were slaughtered.

In many ways, therefore, the Boxer Rebellion of late nineteenth-century China seems to be a reprise of the Great Fear in late eighteenth-century France. In both cases, ENSO-related climate instability contributed to widespread famine and a resulting spike in xenophobic panic. What is more, in both cases, this out-group animus was directed against villainous foreigners and their domestic allies, the aristocrats, in the case of France and the "rice Christians," in the case of China. This is not to say that either the French peasants or Chinese Boxers lacked legitimate grievances against the groups they targeted, or that their actions were unjustified. Nonetheless, the proliferation of unfounded and bizarre rumors in both events suggests that deeper psychological processes may have been involved in the outbreak of both the Great Fear and the Boxer Rebellion, most likely the hyper-activation of the human behavioral immune system during a period of hunger-induced crisis, triggering a wave in xenophobia.

A final way in which the Boxer Rebellion resembled the Great Fear is the role it played in toppling a monarchy. In response to Boxer outrages and military threats, including their siege of the foreign legation quarter in Beijing, a combined expeditionary force of seven European states and Japan captured Beijing in the summer of 1900. In doing so, they matched Boxer atrocities with atrocities of their own, looting, burning, and raping their way through the Chinese countryside, promoting one European observer to quip that "this western civilization of ours is merely a veneer over savagery." In the meantime, Russia exploited the situation to occupy much of Chinese Manchuria, the original homeland of the ruling Qing dynasty. The Qing did not immediately fall in the aftermath of the Boxer Rebellion, in part because the international community still considered the Qing to be the best available alternative to Boxer-style chaos and thus helped to prop up the failing Qing state. Nonetheless, the Qing had been fatally weakened by the Boxer Rebellion, and post-1900 reform attempts only helped to strengthen the hand of anti-monarchical forces, contributing to the founding of the Chinese Republic in 1912 and the abdication of Puyi, the last emperor of the Qing.

Long-Term Climate Change and History

The Science of Radiative Forcing

The ENSO, therefore, has had a marked impact on human history, especially in terms of explaining short-term outbreaks of famine and violence. Other climate processes are equally important in shaping human history, but function on much longer temporal scales. The most crucial such factor that shapes our planet's climate is "radiative forcing," which is a measure of the amount of solar radiation received by the Earth minus the amount of energy radiated back out into space. In general, our climate warms when the input exceeds the output, and cools when the output exceeds the input.

In terms of input, the amount of solar radiation our planet receives varies subtly over the year, the centuries, and the millennia. In geological time, measured in tens of thousands of years, our planet's climate has cooled and warmed due to predictable changes in eccentricity (roundness) of the Earth's orbit, to the Earth's axial tilt, and other factors. Since some of these factors increase radiative forcing, while others decrease it, the Earth's climate at any given point in geological time varies along cycles that range from 25,700 to 100,000 years. The Serbian astronomer Milutin Milankovitch, who first predicted that the Earth's climate might be impacted by slow astrophysical changes, argued in the 1920s that his "Milankovitch Cycles" might explain why the Earth has regularly alternated between ice ages and warm periods, and research has largely borne out his insights. It should be noted that, based on the current interaction of these cycles, our planet should be in the midst of a phase of gradual cooling, though, as we will see below, that cooling is now being counterbalanced by other factors.

A second factor that also impacts the amount of radiation our planet receives is sunspot activity, which varies on the scale of decades rather than millennia. While one might think the Earth receives less radiation when sunspot activity is low, the opposite is true: when many sunspots are observed on the sun's surface, the sun actually puts out slightly more energy than when sunspot activity is weak. The number of observed sunspots varies along an 11-year cycle and may vary along other time intervals as well. During the Maunder Minimum of 1645 to 1715, for example, the number of observed sunspots declined dramatically, and while the cause of this solar event is unknown, we do know that is coincided with a prolonged period of colder European temperatures. Lower sunspot activity and lower planetary temperatures were also noted during the Dalton Minimum of 1789 to 1830. During both periods, however, it is difficult to disentangle the effect of reduced net solar radiation from the impact of major volcanic events, such as Mount Tambora's "year without a summer" in 1816, which occurred during the Dalton Minimum.

While some changes in radiative forcing are the result of changes in input, several other factors can mediate the global temperature by increasing or reducing the Earth's net output of energy. One factor is albedo, in other words, the degree to which our planet's surfaces reflect solar radiation back into space. Clouds reflect about half of the energy of the sun back out to the solar system, so any planetary factor that encourages or suppresses cloud formation can impact radiative forcing and alter the Earth's energy budget. Forests, in turn, have fairly low albedo, while barren soil and sands have higher albedo. As a result, large-scale deforestation or widespread desertification can raise the Earth's albedo and contribute to a reduction of global temperatures, though in the case of deforestation this cooling effect might be counteracted by resulting changes to the chemical composition of the atmosphere, as we will see below.

The effect of albedo on radiative forcing is complex in part because changes to albedo can initiate feedback loops, meaning that even small changes can have big impacts. While the open ocean has low albedo, for example, snow and ice have high albedo, reflecting as much as 90% of the radiation they receive back into space. As a result, if a minor drop in global temperatures contributes to the formation of more sea ice, the resulting reduction of the Earth's albedo can lead to less radiative forcing and further temperature decreases, perhaps initiating a runaway feedback loop and dramatic changes to the Earth's climate in a short period of time. A minor increase in radiative forcing could have the reverse effect, melting ice and reducing the Earth's albedo, leading to runaway global heating.

A final factor that influences the amount of solar radiation retained or lost is the chemical composition of the atmosphere. As we already discussed in Chapter 4 on volcanism, the sulfates released by volcanoes into the upper atmosphere can lead to rapid global cooling by reflecting more radiation back into space. While such cooling is temporary, as sulfates wash out of the atmosphere withing 1–3 years, this short-term chill can be prolonged by feedback loops, in particular, the formation of sea ice and the resultant increase in the Earth's albedo. Volcanic ash may contribute to global cooling as well, though it can also lead to warming if dark-colored ash particles cover ice and snow and lower its surface albedo.

The gases that have the highest impact on radiative forcing, however, are the so-called "greenhouse gases," especially water and carbon dioxide. Most solar radiation is shortwave radiation, and these wavelengths of radiation pass through most atmospheric gases without being absorbed. When they fall on the Earth's surface, they heat it, causing the Earth to emit longwave infrared radiation. Much of this radiation radiates back out to space, but because of their molecular configuration, greenhouse gases capture that energy, heating the atmosphere. While there are natural variations in the amount of these gases in the Earth's atmosphere, human beings have been the main factor behind greenhouse gas-induced radiative forcing since the Industrial Revolution, as we will see in the final section of this chapter.

Climate Change and History in the Ancient World

As a result of the factors outlined above, our planet's climate has experienced centuries-long episodes of relative heating and relative cooling that may have shaped historical development. Between 3200 and 2900 BCE, for example, a prolonged period of weak monsoons called the "Mid-Holocene Crisis" may have had a transformative impact on societies worldwide. According to John L. Brooke (2014), global temperatures dropped during this era, contributing to a persistently negative NAO, the southward retreat of the ITCZ, and an increase in extreme ENSO events, both positive and negative. The result was climate global catastrophes worldwide; as Brooke puts it, "people died in huge numbers, villages were emptied, local gods discredited, ancient memories lost." During this worldwide moment of creative destruction, new societies arose to fill the void, built by charismatic war leaders and stabilized in part by a new caste of literate priests who provided religious justification for the rising regimes. According to Brooke, climate change explains the near-simultaneous rise of complex, hierarchical, and urban societies in Mesopotamia, Egypt, and the Indus valley. While China lagged behind, it was the exception that proved the rule: the monsoon regime remained strong in East Asia until 2400–2000 BCE, around the time of the emergence of China's first dynasty, the Xia.

Even after the end of the "Mid-Holocene Crisis," the climate continued to leave its mark on ancient civilizations. According to Arie S. Issar and Mattanayah Zohar (2007), since the main limiting factor in Middle Eastern agriculture is water availability, there is a close link in this region between the flourishing of agricultural-based states and periods of relative regional humidity. When dryer conditions prevailed, however, the Middle East tended to be overrun with nomadic pastoralists, who respond to drought by seeking new pastures for their flocks, usually at the expense of nearby agriculturalists. For example, the "optimal climate" of mild monsoons in the first half of the third millennium BCE coincided with the Old Kingdom era of Egyptian history, a period of prosperity and pyramid building. In nearby Mesopotamia, this same period was the era of Gilgamesh, a time of urbanization, state-building, and growing cultural sophistication. Between 2400 to 1800 BCE, however, the Middle East suffered from several sustained droughts, contributing to the disintegration of Old Kingdom Egypt and the collapse of the Akkadian Empire in Mesopotamia. More generally, this was a period of de-urbanization throughout the Middle East, and many of the archeological sites in Syria, Anatolia, and Palestine during this period feature layers of ash and embers preserved beneath soils that show few signs of human activity, suggesting a "wave of violent destruction and general upheaval ending all settled life." The same era also features a larger proportion of individual rather than group graves, which is suggestive of a broad shift from agriculture to pastoralism as the predominant lifeway. In contrast, the period from about 1500 to 1200 BCE seems to have been a period of renewed rainfall and urban revival, coinciding with the New Kingdom in Egypt and the unification of Mesopotamia under the Kassite Dynasty.

While much of Issar and Zohar's work is speculative in character, given the relative lack of climate proxies for this era, we are on more solid ground when it comes to the link between climate and the fortunes of the

Roman Empire. To reconstruct the climate of ancient Rome, scientists have relied on proxy data for rainfall and temperature, including tree ring analysis, ice cores, speleothems (cave formations), lake sediments, and historical records, such as annual Nile flood levels in Egypt. Together, this data suggests that the Roman Mediterranean enjoyed an unusually favorable climate during the "Roman Optimum" of 100 BCE to 200 CE, an era of Roman territorial expansion, rapid population growth, and high levels of urbanization. In contrast, the period of 200 CE to 400 CE was one of greater instability, especially during the third century, when a series of volcanic eruptions contributed to rapid cooling, famine, and malnutrition, and repeated outbreaks of epidemic disease. In the meantime, starting in the early fourth century CE, Nile flood levels began to decline, reducing the revenues of Egypt, the breadbasket of Rome. The fourth century also coincided with a severe 40-year drought in the nearby Eurasian Steppe, which McCormick et al. (2012) believe played a "crucial role in driving [the Huns] ... to seek pastures and predation farther to the west and south." The Hunnish advance, in turn, drove the Germanic Goths into Imperial territory, triggering the 378 Roman military defeat at Adrianople, the battlefield death of an Emperor, and an unprecedented crisis for the Roman Empire. McCormick et al. also note that the eastern half of the Roman Empire seems to have enjoyed a more stable and favorable climate than the western half during the fifth century, helping to explain why the Roman Empire collapsed in the west but survived in the east in a Greek-speaking, "Byzantine" form. Taken as a whole, McCormick et al. conclude, "the climate record seems to fit very well with the political/economic history of the Roman Empire."

The Dark Ages Cold Period, the Late Antique Little Ice Age, and the Medieval Warm Period

Just a few generations ago, historians routinely used the term "Dark Ages" to describe the era of European history after the fall of the Western Roman Empire. By the end of the twentieth century, this periodization of history, which had originally been proposed by the Italian Renaissance scholar Petrarch, had fallen from favor. Recent work on climate, however, has suggested that the idea of a "Dark Age" should be revived, though perhaps in a more nuanced way. Numerous paleoclimatic studies indicate that the Northern Hemisphere suffered from a "Dark Ages Cold Period" (DACP) of prolonged overall colder temperatures from about 400 to 800 CE, which most scholars attribute to natural variation in solar radiative forcing. However, the impact and duration of the DACP are hard to distinguish from that of the Late Antique Little Ice Age (LALIA), which as we saw in Chapter 4, began in 536 CE and may have been triggered by three major sulfate-spewing volcanic eruptions in short succession.

In the higher latitudes of Eurasia, declining temperatures due to the DACP (and LALIA) would have posed problems for agriculturalists and pastoralists alike, though in different ways. In farming societies, colder temperatures would have led to declining harvests, population stagnation or decline, and political instability, though societies could try to accommodate to the colder temperatures through agricultural innovation or intensification. Similarly, a shorter growing season would have reduced the growth of grasses upon which pastoralists depend, and as Issar and Zohar have pointed out, the only remedy for bad pastures was more pastures, driving warfare within the Asian steppe and the displacement of societies along the steppe margins. As a result, the DACP has also been called the Migration Period, during which the Germans and Slavs pushed into Europe and Turkish groups invaded China. Thanks to the combined forces of declining crop yields and foreign invasions, Brooke (2014) claims, it would take until 1000 CE for Europe and China to regain the population they had enjoyed in 200 CE at the end of the Roman Optimum.

Nor were the negative impacts of the DACP limited to Eurasia. In the Americas, for example, Brooke argues that "wild extremes" of precipitation between 400 and 1000 CE, linked to "ENSO climatic whiplash conditions," played a role in the collapse of the Teotihuacan, Moche, and Nasca civilizations. At the same time, climate stresses may have helped to spur agricultural innovations in the Americas, such as irrigation systems using high-altitude water sources and the cultivation of crops on raised beds within lakes, where water availability was more reliable. Similarly, in Central America, Mayan city-states that had thrived during the Roman Optimum suffered from recurrent monsoon failures due to the southerly withdrawal of the ITCZ during this era. Like their contemporaries in South America, the Mayans were able to accommodate lower rainfall by means of "elaborate water-management systems," such as wet-season rainfall collection in quarries and lagoons. Nonetheless, classical Mayan civilization would eventually succumb to a series of devastating droughts at the end of the DACP.

At the same time, the DACP period may have been of some benefit to societies in the Middle East. As we saw in Chapter 4, the proliferation of scrub vegetation during this cool snap may have facilitated the Arab conquests. Büntgen et al. (2016) link this to the LALIA, but the same phenomenon could just as easily be ascribed to the DACP, which seems to have locked the NAO climate system into a prevailing negative phase, leading to persistent cold and dry conditions in northern Europe but relatively wet weather in the Mediterranean and parts of the Middle East, including the Anatolian highlands that are the headwaters of the Tigris and Euphrates Rivers. Whatever the cause, the Islamic Caliphates seem to have benefitted from a wet climate regime, which Brooke (2014) speculates "may have provided the material underpinnings of the strength of the Islamic Golden Age." This climatic interpretation of Islam's heyday is not universally accepted, however. In any case, the new crops that Arab traders imported from India during this period, which, according to Andrew Watson (1983), included sorghum, sugar, cotton, spinach, and citrus fruits, probably also played a part in nourishing the Golden Age of Islam.

But what climate giveth, climate can taketh away. By around 950 CE, the Dark Ages Cold Period was giving way to the Medieval Warm Period (MWP), during which increased radiative forcing and decreased volcanic activity caused temperatures in the Northern Hemisphere to tick upwards for three centuries. In the process, the NAO switched from a prevailing negative to a generally positive mode, drying up the stream of moisture from the Atlantic Ocean to the Middle East that had previously contributed to the prosperity in the era of the Islamic Caliphates. As a result, according to Ronnie Ellenblum (2012), the Middle East experienced the beginning of the MWP as a series of serious droughts, first in 1004–1009 and again in the 1020s. The Arabian Peninsula in particular suffered from hyperarid conditions during the MWP, and Baghdad's prosperity was undermined by lower flow rates in the Tigris and Euphrates, which depended on winter-season moisture blown in from the Atlantic. What is more, although the exact climatic cause is unclear, Egypt experienced extremely low Nile water levels during this period, leading to famines "of biblical dimensions" in the 1040s and 1070s, the latter of which helped to bring down the Fatimid Dynasty.

According to Richard Bulliet (2009), the Middle East's problems during the Medieval Warm Period were particularly acute in Persia. Paradoxically, Persia's temperatures seem to have dropped during the Medieval "Warm" Period, cutting short an earlier ninth- to tenth-century boom in Persian cotton production. What is more, the "Big Chill" (as Bulliet calls it) seems to have also lowered temperatures in the nearby Central Asian steppe, shortening the grass-growing season. This combination of weakness in Persia and crisis on the steppe contributed to the widespread infiltration of Turkish pastoralists into Persia during this period. The Turks initially were imported into Persia as military slaves (*mamluks*), but as is often the case in the Islamic Middle East, the slaves seized power for themselves, founding a dynasty of Turkish origins, the Ghaznavids. The Ghaznavids were themselves overthrown by invading Seljuk Turks in the early eleventh century, whose own Seljuk dynasty was threatened in turn by new waves of nomadic invaders from the steppes, including the Mongols, who swept away what was left of the Seljuk domains in the mid-thirteenth century. One of the legacies of Iran's troubled history in this period, Bulliet claims, was the flight of Persian scholars to Syria, Egypt, Anatolia, and Egypt. To at least some extent, he argues, these Persian men of letters can be considered climate refugees, as were the migrating pastoralist tribesmen who helped to displace them.

Other global societies suffered adverse climate effects during the Medieval Warm Period as well. In a study, for example, Cook et al. (2007) used 835 tree ring sequences spread over the USA, Canada, and Mexico to connect weather to a series of societal collapses in the North America, both during the MWP and afterwards. According to these authors, the period of 900–1300, which roughly coincides with the MWP, was generally dry in North America, especially in the west. What is more, these prevailing arid conditions were occasionally punctuated by severe drought events, including the "megadrought" of 1140–1162, a period of 23 consecutive rainfall deficit years.

In Europe, however, the Medieval Warm Period was an unmitigated blessing. The same consistently positive NAO that dried out the Islamic world brought several centuries of warm and fairly stable climatic conditions to Northern Europe, extending the growing season in much of Europe by three weeks over what was possible during the DACP. As a result, Brian Fagan (2008) notes that Europe's "growing population moved northward and uphill," colonizing high latitudes and high elevations that had previously been marginal for agriculture. Iceland, for example, was first settled by Vikings in 865 and 930, at around the time that the DACP was coming to an end, and Greenland was colonized around 1000, when the MWP was well underway. In some parts of highland Europe, crops could now be planted and animals grazed on land that had been covered by glaciers just two centuries before. More broadly, Europe's populations rose dramatically in this era, rising from about 35 million in 1000 CE to 80 million in 1347, on the eve of the Black Death. As Fagan points out, Europe's imposing Gothic cathedrals, many of which date to this period, are a tribute to the high populations and surplus resources of the MWP.

In China, the story of the MWP is also one of demographic growth, though in this case the gains of the WMP were largely reversed by destructive warfare against invading pastoralists from the Asian steppes. According to Brooke (2014), the Chinese population grew from roughly 56 million in 1000 CE to fully 128 million in 1200, bolstered by warm summer monsoons as well as the cultivation of *champa* rice, a quick-growing east Indian variety introduced to China in the eleventh century that can support multiple harvests per year. Unfortunately for China, the MWP was an era of aridity in the Asian steppes, which may have helped to impel first the Manchurian Jurchen and later the Mongols into Chinese territory. In particular, the "insanely destructive" Mongol invasion seems to have undone much of the demographic gains China had made during the MWP; Brooke notes that northern China may have lost over 80% of its population from 1195 to 1235, dropping from 50 million to just 8.5 million. As a result, China's population during the Mongol Yuan dynasty was only about half of what it had been during the height of the MWP.

The Little Ice Age

Of all the climatic eras considered in this chapter, the Little Ice Age (LIA) is perhaps the most studied, but the least understood. For one thing, scientists do not agree on what caused this sustained drop in Northern

Hemisphere temperatures. A reduction in solar radiation intensity may have played a role, and indeed the worst years of the LIA overlapped with the Maunder Minimum, a sustained period of low sunspot activity. Other studies, such as an article by Gifford Miller et al. (2012), argue that volcanism served as the initial trigger. Cooling caused by volcanic sulfates in the 1270s, they contend, initiated a feedback mechanism during which volcanic winters triggered pack ice formation in the Arctic, leading to a higher global surface albedo that further cooled the planet and stimulated still more ice formation. The most intriguing theory about the origins of the LIA is that it was an unexpected side effect of disease outbreaks, specifically the Black Death epidemic of 1346–1351, which carried away as much as a third of the population in parts of the Old World, and the Columbian Exchange epidemics of the sixteenth century, which would reduce the Amerindian population of the New World by as much as 95%. We will return to this last point in the final section of this chapter, which considers anthropogenic causes of climate change.

Nor is there any agreement on when the LIA began or ended. Some scholars would push the LIA as far back as the mid-thirteenth century, when pack ice began to build up in the North Atlantic and glaciers began to form in the Arctic. Others prefer to date it to the catastrophic 1315–1317 European famine, which killed an estimated 5–12% of Europe's population. Complicating matters further, Brooke (2014) proposes that the period from the late thirteenth to the late nineteenth century should be divided into *three* distinct periods: (1) a transitional era of variable weather from about 1150–1400 CE; (2) a "LIA I" of cold winters from 1400–1550; and (3) a "LIA II" of colder winters and summers from 1550–1725. David Zhang et al. (2011) also propose a tripartite division of the LIA, though a different one: (1) a mild phase from 1500–1559; (2) a cold snap from 1560–1660; and (3) then a second mild phase from 1661–1800. As for its end date, different scholars would date it to 1800, 1850, or even to the early twentieth century.

Whatever the exact chronology, overall temperatures in the Northern Hemisphere were significantly below those of the Medieval Warm Period. The growing season was shortened in some areas by as much as five weeks, and winters grew colder and stormier. After nearly five centuries of continuous occupation, Greenland was abandoned entirely by Norse settlers in the mid-sixteenth century, and at around the same time, glaciers began to advance throughout the highlands of Europe, reclaiming the land they had surrendered to crops and livestock during the MWP. Wine grapes, which had been planted as far north as England and southern Norway in the benign climate of the MWP, retreated back to the south. What is more, harvest failures became increasingly common, leading to higher food prices, hunger, and occasional famine. At least during the first half of the Little Ice Age, Brian Fagan (2000) claims, "hunger was never far from the door."

If Wolfgang Behringer (1999) is correct, the climate setbacks of the Little Ice Age may help to explain one of the most notorious phenomena of the early modern era: the European witchcraft trials. Just as Europeans had blamed Jews for the Black Death, they now sought someone to blame for the deteriorating weather and found that scapegoat in the "weather-working witch," usually female, who was increasingly blamed for poor harvests. As Brooke (2014) points out, early modern Europe was by no means unique in this regard. Witchcraft accusations also broke out in China during the LIA, as well as across the Atlantic, where a "witch panic … gripped the Hurons of North America between 1635 and 1645." In terms of scale and intensity, however, Europe's witchcraft hysteria was in a category of its own.

The first systematic witch hunts took place in Savoy in the 1430s, but for the next century these persecutions were localized to the "Italian, French, and Swiss Alpine valleys" – marginal areas which would have been first impacted by the cooling weather of the LIA. As the climate worsened further after 1550, however, popular calls for witchcraft prosecutions grew in frequency and urgency. Things came to a head in southern Germany in the early 1560s, following a catalogue of climate calamities. The years 1560 and 1561 experienced unusually wet summers and frigid winters, so cold that Lake Constance in the nearby Alps froze for the first time in living memory. Then, in 1562, heavy spring and early summer rainfall brought flooding and cattle diseases to the region, while late summer witnessed terrifying thunderstorms and damaging hail. Conceding to popular "demonstrations and petitions of the peasants," local leaders imprisoned some women suspected of weather-making, those women (under torture) implicated other women, and the witch hunts snowballed from there. While the small states of southern Germany experienced the most witch trials and the most executions, which Behringer ascribes to the inability of weak local governments to resist popular pressure, similar trials occurred elsewhere, including England, Scotland, France, Switzerland, Hungary, Austria, and Poland. As many as 60,000 accused witches were judicially murdered before the witch-hunting mania petered out after 1750. Some 40,000 were killed during the period of 1560–1630 alone, years that overlap quite well with the nadir of the LIA.

As Behringer (1999) points out, the "synchronicity of [witchcraft] accusations and persecutions in these far-away countries, not connected by dynastic, confessional, economic, or other links," suggests a common causative factor, which he believes was the Little Ice Age's deteriorating climate. Each pulse of bad weather, he notes, created "enormous psychological stress" among contemporaries, which manifested itself in anti-witch hysteria that Behringer compares to the Great Fear of the French Revolution. The likely psychological mechanism that turned fear and deprivation into scapegoating and massacre was, of course, the behavioral immune system, which as we have seen repeatedly over the course of this book, can become hypersensitive during periods of hunger and perceived disease vulnerability, triggering out-group animus.

But the BIS did not act alone. In the decades before 1560, the ground was prepared for the witch-hunts by popular anti-witchcraft tracts, such as the Catholic clergyman Heinrich Kramer's *Malleus Maleficarum*. Published in 1486, Kramer's "Hammer of Witches" accused witches of causing hailstorms and other disasters with their wicked spells, and his claims reached a wide audience thanks to the recently invented Gutenberg printing press. Even the illiterate were convinced of the dangers of witchy weather-working thanks to mass-printed woodcuts depicting spell-casting witches flying over thunderstorms or conjuring hailstorms with their devilish sacrifices (Figure 5.3). Such accusations may also have been driven by misogyny. Kramer was typical of many Europeans of the age in believing that the "weaker" sex was more susceptible to the temptations of the Devil. In addition, population expansion during the Medieval Warm Period had created land shortages and resultant resentment against older land-holding widows, whose very existence was blocking the marital aspirations of younger men. As a result, by the time that the Little Ice Age reached its chilling climax, both cultural and socioeconomic factors helped to focus xenophobic hostility against the female weather-working witch. Witchcraft accusations would only die out in the eighteenth century, in part due to the growing rationalism of the Enlightenment Era, though it is likely that improving weather conditions played a role as well.

Weather downturns during the same period also left an ugly mark on Russian history by contributing to the "Time of Troubles" of 1589–1613. According to Chester Dunning (1995), just as Russia "appeared to be emerging as a powerful state capable of challenging its European neighbors on equal terms," a "catastrophic internal crisis" marked by famines and bloody civil wars "nearly destroyed the country." While many historians have blamed the Time of Troubles on class tensions, seeing it as "a social struggle of the masses against the development of serfdom," Dunning argues for a more complex suite of causative factors, including the Little Ice Age, which further narrowed an already-short Russian growing season. As a consequence, the 1601–1603 famine was so severe that it led to widespread reports of cannibalism as well as the death of up to a third of Russia's population. Not only did this famine destabilize the regime of Tzar Boris Godunov, it delegitimized Godunov's rule, since Russians believed that natural disasters were signs of divine displeasure with a bad Tzar. It should be noted that the extraordinarily low temperatures that contributed to the Time of Troubles famine may be the result of the combined impact of the LIA and the sulfates of Peru's Huaynaputina Volcano, which exploded in 1600 and seems to have triggered a cold snap throughout the Northern Hemisphere.

As Sam White (2011) has chronicled in his excellent book on the Celali Rebellions, the "troubles" that marked the end of the sixteenth century

Figure 5.3 Weather-working witches conjuring a hailstorm during the Little Ice Age
Source: Universal History Archive/Getty.

were by no means confined to Russia. According to White, the Ottoman Empire was particularly vulnerable to climatic down-turns by the late 1500s. Thanks to impressive demographic expansion during the milder early phase of the LIA, the inhabitants of Ottoman Anatolia had expanded into ever more marginal lands, though the supply of such lands did not come close to matching population growth, creating a dangerous surplus of landless young men. What is more, the Ottoman core territory of Asia Minor hosted large numbers of Turkmen and other pastoralists, who, as we saw repeatedly earlier in this chapter, tend to react to climate setbacks by aggressively seeking to expand their territories. These building social tensions were exacerbated by a LIA-linked drought, which began in the early 1590s. The drought peaked in 1596, at around the same time the Ottoman government requisitioned a huge tribute in livestock from the already-starving inhabitants of the Karaman province of interior Anatolia to help fund a renewed round of warfare with the Austrian Habsburgs.

The resulting rebellions, White tells us, precipitated "the worst crisis in Ottoman history from the invasion of Tamerlane to World War I." Rural banditry broke out in the parched Anatolian interior, and when these bandit groups began to coalesce under the leadership of charismatic captains, what started as desperate lawlessness morphed into outright rebellion against the Ottoman state. In the meantime, many of the rural peasants of the Anatolian interior fled to the urban centers, where by and large they died, falling prey to hunger as well as big-city diseases and the effects of overcrowding. Worse of all, much of the abandoned countryside was taken over by Turkmen tribes and other pastoralist groups, creating a long-standing military threat to the state while simultaneously depriving it of tax revenues. While the Ottoman state would eventually get the Celali Rebellions under control, White argues that the Celali rebellion and subsequent LIA-linked famines in the seventeenth century greatly weakened the Ottoman Empire vis-à-vis its European competitors. White does not deny the importance of non-climate factors in driving the events of this era; the Turkmen invasion of Anatolia, for example, was driven in part by aggressive military campaigns by the Persian Safavid dynasty, which displaced many Turkmen tribes westward into Ottoman territory. Nonetheless, White concludes that "climate was a crucial factor – perhaps *the* crucial factor – in understanding the Ottoman crises" of the LIA.

The "Global Crisis" of the Seventeenth Century

According to Geoffrey Parker (2013), these Russian and Ottoman crises at the end of the sixteenth century were just the prelude to the main show, the dramatic unfolding of the calamitous seventeenth century. Some twentieth-century historians, especially Eric Hobsbawm and Hugh Trevor-Roper (1959), had noted the simultaneous outbreak of political crises during the seventeenth century throughout Western and Central Europe, which Trevor-Roper ascribed to a general pattern of popular resistance against state centralization. In essence, he contended, these crises were nothing less than the growing pains of the modern nation-state. While not denying the validity of some of Trevor-Roper's conclusions, Parker argued that the concept of the seventeenth-century crisis needs to be widened and revised: widened to include the whole planet, and revised to incorporate what we now know about climate changes during the Little Ice Age, which bottomed out in the early and middle years of the seventeenth century.

For Parker, the seventeenth century was a true "global crisis," during which "intense … global cooling" triggered an "unparalleled spate of revolutions and state breakdowns around the world." Between 1620 to 1680, the years where Parker (2013) focuses his attention, the monsoons weakened and the growing season shortened, to the point that the chance of two harvest failures occurring back-to-back quadrupled. As a result of the frequent famines of this period, Parker notes, "French soldiers born in the second half of the seventeenth century were on average about an inch shorter than those born after 1700; and those born in famine years were notably shorter than the rest." Based on the Quebec Ice Storm study findings we discussed in Chapter 1, it is likely that shortness of stature was far from the only problem faced by this famine-era age cohort: the French of the seventeenth century probably suffered from a whole spectrum of other developmental disorders, including low IQ, language problems, and high rates of diabetes. The climate woes of this period also triggered both epidemic disease and equally epidemic warfare. "More wars took place around the world," Parker claims, "than in any other era before the Second World War."

Nonetheless, Parker notes, the seventeenth-century crisis unfolded in different regions in distinct ways, underscoring the importance of considering local circumstances and cultural variations when assessing the outcomes of disastrous events. A cruel constellation of factors, for example, ensured that Ming China suffered "worst and longest" during the global crisis. The Ming were already in a state of semi-crisis before the weather turned sour, since in-fighting between different cliques in the bureaucracy had weakened the authority of the central government, and the ineffective emperors of the day sought only "instant successes and simple solutions" to China's increasingly complex problems. After a series of monsoon failures in the 1620s, banditry spiked in the afflicted regions, and the White Lotus secret society organized a rebellion in Shandong, the same province that would nurture the Boxers at the end of the nineteenth century. Due to irregular salary payments during this period, soldiers sent to suppress the bandits often deserted to join them instead, bolstering their numbers. By the 1630s, the weather deteriorated further, and "petitions flowed in to the central government from all over the Empire, begging for action to end the bandit menace and relieve the suffering caused by failed harvests, high taxes, and bad weather." As in the age of the Late Victorian Holocausts discussed above, famine struck Shanxi and Shandong in China's north, and grim rumors circulated of "fathers and children, siblings and husbands resort[ing] to cannibalism."

Worst of all, the Ming were distracted from these internal threats by an external menace, the Manchus, who had broken through the Great Wall and were ravaging northern China. To some degree, the Manchu invasion was itself a response to the period's extreme climate challenges: the Manchu homeland was badly impacted by the region's cold weather, so the initial Manchu war objective was the agriculturally-rich province of Liaodong in northern China, in hopes that its harvests could be used to alleviate Manchuria's famine. The Ming armies sent to counter the Manchu threat proved to be unequal to the task, in part because of widespread corruption. Parker notes, for example, that Chinese army inspectors of the day lamented:

> If 100,000 names appeared on the army's [muster roles], only 50,000 soldiers actually served … [and] of those 50,000 men, no more than half are any use in combat. The Court thus pays for four soldiers but receives the services of only one.

In the meantime, the Manchu army was becoming more formidable thanks to its adoption of Chinese artillery and musket technology.

The breaking point for the Ming Dynasty finally came in the early 1640s, when the already-cold weather was further cooled by a series of major volcanic events, including the nearby eruption of Japan's Komagatake volcano in 1640, and the ENSO cycle shifted into an extreme El Niño mode. As a result, "North China experienced the worst drought recorded during the last five centuries," and since parts of the Grand Canal dried up due to lack of river flow, hunger in the north could not be alleviated with the rice surpluses of the south. In any case, southern China was facing its own problems as well, since a shortage of silver coinage wreaked havoc on the South's cash-based economy. Focused on beating the Manchu, the Ming essentially abdicated responsibility for famine relief, leading to a further reports of cannibalism and a rise in banditry.

As was the case in the Ottoman Empire a generation earlier, many of these bandit gangs began to coalesce into armies under the leadership of charismatic warlords, including Li Zicheng, a former postal carrier of humble birth known to his followers as the "Dashing Prince." By 1642, Li was strong enough to capture the former imperial capital of Kaifeng, and in 1643 he captured the Ming capital in Beijing, prompting the suicide of the reigning emperor. However, there was one more remaining obstacle to the imperial claims of the Dashing Prince: the army of Wu Sanghui, a Ming general whose men were guarding the strategic Shanhai Pass against Manchu invasion from the north. When threatened by the larger force of the Dashing Prince, Wu made the fateful decision to appeal to the Manchu for assistance. The Manchus exploited the situation skillfully, allowing Li to deplete Wu's army before intervening and forcing Li to retreat, after which they compelled Wu and his soldiers to accept vassalage under the Manchu. Soon after, this combined Manchu/Chinese "Grand Army" seized Beijing from the Dashing Prince and enthroned the Manchu king as the "Son of Heaven," marking the start of the Qing dynasty.

While the victory of the Manchus was due in part due to divisions among the Chinese, it also reflected the reckless desperation of the Manchus in an age of deteriorating climate. The total population of the Manchus was only about a million in the mid-seventeenth century, so the

60,000-strong Manchu force that seized Beijing represented a significant fraction of all male Manchu adults, giving the expedition more of the character of a mass migration than a military campaign. Once Beijing was seized and the Qing dynasty proclaimed, many more Manchu fled the cold and famine-prone north for China proper, which helps to explain how the Qing were able to follow up on their initial successes and impose Manchu control over China despite their low numbers. As a result, David Zheng et al. (2007) argue, climate change provides the best explanation for how the greatly outnumbered Manchus could "swallow an enormous empire with profound cultural heritage." Luckily for the Qing, by the time they had subdued the Ming loyalists of Southern China by the early 1680s, China's climate had begun to improve. In any case, political unification allowed the Qing to marshal China's vast national resources to combat regional famine, solidifying Qing authority in the process.

According to David Zheng and his colleagues, the fall of the Ming in time of climate change is characteristic of a general tendency in Chinese history. In a pair of sweeping studies, Zheng et al. (2006; 2007) argued that "during cold phases, China suffered more often from frequent wars, population decline and dynastic changes," and noted that the collapse of the last three dynasties in China – the Song, Ming, and Qing – all coincided with cold snaps. While South China seems to have been somewhat resistant to climate shocks, their study found a "near-perfect match" between climate change and warfare in the north. Like Issar and Zohar (2007), they ascribe this tendency to the realities of pastoralism, which is far more prevalent in China's dryer and temperate north than its subtropical south. A drop in annual average temperatures of 2°C, for example, can "shorten the grass growing season by 40 days" in Manchuria and Northern China, which would have impelled nomadic pastoralists into Chinese territory in search of better grazing. Overall, Zheng et al. suggest that climate change provides a much better explanation for the rise and fall of China's dynastic cycles than traditional Chinese historiography, which ascribes the success or failure of dynasties to "social evolution or internal management" rather than climatic factors.

While China may have suffered in a particularly dramatic fashion from the seventeenth-century crisis, China's traumas were echoed worldwide. Nonetheless, as Parker (2013) demonstrates in his exhaustively-researched book, each of the seventeenth-century crises played out in different political, cultural, and other contexts, leading to divergent outcomes. The Mughal state of India, for example, suffered from the same monsoon failures as China in the 1630s, leading to widespread famine, but India's famines did not lead to regime change in part because the ruling Shah "used some of the vast resources at his disposal to respond to the disaster." As for Japan, it not only survived the LIA, it prospered during the seventeenth century, rising in population from 12 million at the start of the seventeenth century to 27 million by its end. This is not because Japan avoided famine. The cold winter of

1641–1642 shortened the growing season in Japan by up to six weeks and triggered widespread hunger, just as it did in China during the same period. Overall, however, Parker argues that Tokugawa Japan was one part of the world that "got it right" during the Little Ice Age, finding ways to mitigate its impacts. Japan's ability to flourish during a climate downturn is all the more impressive given the fact that, unlike the states of Western Europe on the other side of Eurasian continent, Japan did not have an American safety valve to drain off its excess population, nor could it draw upon the vast resources of the Americas to compensate for LIA-era shortages.

So what made Japan different from China, Russia, and the Ottomans? For one thing, Japan entered the seventeenth century with a comparatively low population, having just recovered from a bloody unification war and failed military campaigns against Korea. Drawing on the work of William Atwell (1986) and other authors, Parker (2013) also ascribes Japan's stability to the highly centralized and ruthlessly efficient Tokugawa state. When famine struck in 1641–1642, for example, the Tokugawa government "set up food kitchens and shelters for the starving," released rice from government granaries, restricted the labor demands that the *daimyō* (feudal lords) could require from the peasants, and lowered state tax demands, in hopes that this would make foodstuffs more affordable. When prices continued to climb despite these measures, the Tokugawa forbade the use of rice to make *sake* and ordered the execution of grain speculators who were withholding stocks from market. Such state intervention, Parker claims, allowed Japan to escape both the widespread famine and the large-scale rebellions that plagued contemporary states during the seventeenth-century crisis.

A final factor in Tokugawa success was the shogunate's "risk-adverse foreign policy." While Tokugawa Japan did provide asylum to some Ming loyalists who had fled Qing-controlled China, the shoguns overall "scrupulously avoided foreign intervention." In the long run, Parker notes, this policy might have inhibited military innovation in Japan, and Japan's resultant inability to resist the intrusion of the technologically superior West in the mid-nineteenth century helped to topple the Tokugawa Shogunate. In the short and medium run, however, the Tokugawa state's isolationism kept it out of the wars that "drained the revenues of most other early modern states," allowing it to "keep tax rates relatively low and yet still accumulate resources with which to respond effectively in case of a natural disaster." In the end, Japan's experience during the Little Ice Age illustrates that local contexts do matter and that efficient government action can lessen the impact of even a global-scale disaster – a hopeful lesson for us all, as the world careens into the modern age of human-induced climate change.

Anthropogenic Climate Change and Global Warming

As we saw earlier in this chapter, the world's climate is governed in large part by radiative forcing, the level of which can be altered either by increasing/decreasing the input of energy (e.g., during sunspot minimums) or increasing/decreasing the solar energy lost to the void. Needless to say, humans have little influence on the "input" side of the equation. We have no dimmer switch for the sun, though some people have suggested that giant space mirrors capable of intercepting solar radiation might be a viable way to reduce global warming. However, human beings can and do impact the "output" of energy back into space, potentially altering the planet's climate in the process. We do this in two main ways: (1) by changing the Earth's albedo; and (2) by tinkering with the chemical composition in our atmosphere, particularly through the release of greenhouse gases and sulfates.

While many of us associate human-induced climate change with the Industrial Revolution and modernity, paleoclimatologist William Ruddiman (2005) contends that anthropogenic climate change began much earlier. In the early 2000s, Ruddiman noticed an apparent discrepancy between simulated climate projections and climate actual data. While global CO_2 levels should have been dropping over the last 10,000 years due to the natural workings of the Milankovitch Cycles, Ruddiman observed that CO_2 levels have stayed steady or even risen during that period. The solution to the riddle, Ruddiman contends, is human agriculture. When early farmers cut down trees or burned shrubs to prepare the soil for planting, they released small but cumulatively significant amounts of CO_2 into the atmosphere. Similarly, rice paddy agriculture increased methane into the atmosphere, a short-lived but potent greenhouse gas. Ruddiman therefore argues that the start of the "Anthropocene" – the modern-day era of human-induced climate change – should date back, not to the Industrial Revolution, but to the dawn of agriculture. If not for the intervention of agriculture, Ruddiman contends, the world would now be well on the way to another ice age.

More recent research into Ruddiman's hypothesis suggests that agriculture might have initiated a positive feedback loop, in which an increase in farmed land created climatic conditions favorable to farming, setting the stage for yet more agricultural production. A study by Stephen Vavrus et al. in 2022, for example, has argued that the surface albedo and greenhouse emissions changes wrought by ancient farming made the world overall warmer, wetter, and overall better for agricultural production. This was particularly true in higher latitude areas in North America and Eurasia, which (according to the authors) may help to explain why the center of agricultural production shifted northward over time, from its original cradle in Mesopotamia and the Fertile Crescent into Europe, China, and elsewhere.

On the flip side of this, Ruddiman (2005) argues that historical population crashes have triggered global cooling, creating possible feedback loops that can render the world *less* suitable for agriculture. The link between climate and population, Ruddiman claims, is global-scale disease outbreaks. In the wake of major plagues, reduced food demand means that marginal lands are abandoned to nature, and reduced human populations put less pressure on forest resources, contributing to reforestation. In the process, megatons of atmospheric carbon are "sequestered" (withdrawn) from the atmosphere. While we might think that trees grow from the ground up, they are for the most part built out of thin air. Wood is 50% carbon by volume, which plants obtain by absorbing the greenhouse gas CO_2 from the atmosphere themselves.

Whenever large-scale reforestation occurs due to disease, therefore, the climate consequences can be dramatic. The bubonic plague of the early fourteenth century killed so many people in Eurasia and Africa, Ruddiman contends, that the resulting reforestation and carbon sequestration jumpstarted the Little Ice Age. Even greater episodes of reforestation occurred in the fifteenth and sixteenth centuries due to the depopulation of the Americas as a result of the "Columbian Exchange," which introduced Old World pathogens to immunologically defenseless New World societies, reducing the population of the Americas by as much as 95%. Small wonder, Ruddiman contends, that the nadir of the Little Ice Age occurred during the period from approximately 1500 to 1750 CE, when Amerindian populations were declining and had not yet been replaced by new inhabitants of European and African ancestry. Brooke (2014) has made the interesting suggestion that the Columbian Exchange may also have contributed to the LIA through its impacts in Africa, since the growth of the African slave trade to feed European demands set off cycles of warfare, depopulation, and resulting reforestation. As Brooke points out, however, no scholar has yet "attempted to calculate the CO_2 impact of this decline."

From Global Cooling to Global Warming: Climate Change Today and Tomorrow

While it is debatable whether anthropogenic factors contributed to the Little Ice Age, it is almost universally accepted that greenhouse gas emissions in the age of industry are what brought the LIA to an eventual end. At the start of the nineteenth century, the concentration of CO_2 in the world's atmosphere stood at 275 parts per million, and this level had been fairly stable since about 1 CE. CO_2 levels in the atmosphere rose sharply after 1800, however, climbing to about 300 ppm in 1900, 370 ppm in 2000, and just below 420 ppm as I write this in late 2023. The main driver in this has been the burning of fossil hydrocarbons, ancient plant matter buried underground for millions of years and transmuted into combustible minerals by heat and pressure. Coal began to be burned industrially on a large scale during the "first industrial revolution" of 1760–1840, especially after 1800, when steam engines came into widespread use. The heyday of oil, in turn, began during the "second industrial revolution" following the invention of the internal combustion engine, and peaked in the automotive era, which began in the early twentieth century and continues until today.

In some ways, the human use of fossil fuels during the modern industrial era does not represent a dramatic break from the past. Although people don't usually think of wood as a hydrocarbon, about 90% of wood consists of hydrogen and carbon atoms, so the difference between wood and coal is more about vintage than substance. Many early steam trains of the early industrial era ran on wood, especially in America, which still had substantial timber reserves at the start of the nineteenth century. By the same time, however, Europe's timber stocks were becoming increasingly meager, forcing the British and other Europeans to progressively switch over from wood to coal.

While Europe may have turned to coal out of necessity as much as by choice, coal did have some advantages over wood. For one thing, coal has about 25% more energy density than wood, making it a better option for vehicles like trains and steamships that have to carry their own fuel supply in addition to passengers and cargo. Oil, and its byproducts diesel and gasoline, have an even greater energy density. More importantly, while wood is a renewable resource, it is also a limited one, both because the amount of forest acreage a country possesses is finite and because trees take a generation or more to reach a harvestable height, which limits the total energy available annually. Fossil fuels, in contrast, are not renewable – once they are gone, they are gone – but at least in the short run, they can be enormously productive. A single coal vein might contain the solar energy captured by an ancient forest over the course of thousands of years, so industrialized societies are essentially burning the candle at both ends, exploiting both the solar energy of the present day as well as the captured solar energy of the distant past.

Fossil fuels, therefore, allowed Europeans of the industrial era to transcend the limitations of the old solar economy by mining for buried sunlight. Unfortunately, since the burning fossil fuels has also released long-buried carbon back out into the world, the result has been a steady climb in global temperatures which tracks rising CO_2 levels with near-perfect fidelity. Average global temperatures have risen by .08° per decade since 1880, and rate of temperature increases is now accelerating, increasing by .18° every ten years. While temperature increases of a fraction of a degree might not seem like much, Anthony McMichael (2017) notes that the world is now growing warmer *thirty times faster* than it did at the end of the last ice age. If we stay on our current trajectory, Kemp et al. (2022) project that global temperatures will be 2.1°C–3.9°C higher in 2100 than they are today, reaching temperature levels that the world hasn't seen since the Piacenzian Age (3.6–2.59 million years ago), a time of mass extinctions and sea levels that were 20 meters higher than those of the present day. The country where this book was written, Qatar, did not even exist during most of the Piacenzian, being completely submerged beneath the waters of a much-wider Persian Gulf.

Can human civilization, as we know it, survive such a sharp increase in temperatures over such a short span of time? Some scholars believe it cannot. Australian epidemiologist Anthony McMichael (2017) has predicted that global warming will create an "environmentally disrupted, resource-depleted, and conflict-prone world," in which privileged minority enclaves occupying lands less affected by rising temperatures will defend their borders and their advantages in a world rendered largely unhabitable by climate change. Economics professor John Gowdy (2020) has gone even farther, predicting a future world of "uncivilization," where agriculture will become impossible amidst soaring temperatures and humanity will be obliged to return to hunter-gathering for survival.

Other scholars are more optimistic about humanity's future. In 2012, archeologist and climatologist Karl Butzer lamented the rising "alarmist" tendency in academia and noted that history records plenty of successful examples of adaptation to climate change, including the ability of much of Europe to survive and even flourish during the Little Ice Age. In a similar vein, climate historian James Fleming (2014) has criticized the recent rise of "prophets of climate catastrophe" who see in every unusual weather event a "pending dystopian future that can only be avoided if humanity collectively, immediately, and dramatically changes its ways." These academics, Fleming claims, have wandered from scholarship into "eschatology, the doctrine of the end of the world; and it is here that climatology meets theology." What is more, while Fleming certainly recognizes the reality of global warming and denounces climate deniers, he is equally critical of "dangerous rogue groups of climate engineers" who "propose to 'fix' the climate with heavy-handed technologies."

It is still far too early to make broad predictions about the effect that global warming will have on our collective human future. We are on more solid ground when we consider the impact that global warming is having on the present day, especially in ecologically and socioeconomically vulnerable locations that are particularly sensitive to climate change. In some places of the world, global warming is already making history.

Societies located in the higher latitudes of the Northern Hemisphere, for example, are already experiencing substantial temperature increases and corresponding economic hardship. Indigenous communities in the Arctic Circle like the Nets'aii Gwich'in of north central Alaska, for example, are contending with 1–3°C temperature increases, leading to greater winter snowfall, but also thinner ice, which complicates transportation between towns and to hunting grounds. According to anthropologist Steven C. Dinero (2016), the Nets'aii Gwich'in have also noted that caribou herds are thinning, in both senses of the word – not only are caribou becoming fewer in number (or at least more difficult to obtain), their body fat percentage seems to be dropping, reducing the quality of the meat. In addition, the region's permafrost soils are thawing, and as a result, some lakes are "caving in" or drying up, including many favored fishing locations. Dinero notes that Alaskan Arctic populations have so far proven to be resilient in the face of these changes. Nonetheless, since Alaska is warming at twice the rate of the continental United States, the ability of the Nets'aii Gwich'in to maintain traditional lifeways will be put to a severe test in coming decades.

Substantial recent temperature increases have also been reported in Scandinavia, where the growing season is expected to grow by a full month by 2050 and two months by the end of the century. While a longer growing season might sound advantageous to local agriculture, the extreme drought of 2018 suggests otherwise: temperatures that year in Eastern Norway were 4.3°C above the 1961–1990 average, and according to Bjørnar Sæther and Karen O'Brien (2022), low rainfall levels combined with extreme soil desiccation led to a 50% drop in overall grain production. In many places, due to lack of soil moisture, the grain never even sprouted, much less grew to maturity. Other recent years in Norway, on the other hand, have been anomalously wet, leading to crop losses as well as the risk of "increased runoff, nutrient loss, and soil erosion, as well as soil compaction by heavy machinery." Nonetheless, like the Nets'aii Gwich'in, Norwegian farmers are adapting to these new challenges, for instance, through construction of new drainage ditches to counter the impact of increasing rainfall. They have also tried to mitigate resource losses, in particular lack of animal fodder, through cooperative organizations and self-help societies.

Moving from the frozen (but thawing) north to the balmy subtropics, the South Asian nation of Bangladesh represents a second front line of climate change in the modern world. The primary way that climate change has impacted Bangladesh is through flooding. The Himalayan Range's glaciers are melting faster due to global warming, contributing to raging downstream floods and widespread riverbank erosion; Tim McDonnell reported in 2019 that "every year" in Bangladesh, "an area larger than Manhattan washes away." In the meantime, due to ice melt and thermal expansion, sea levels are rising worldwide, which is a serious problem in Bangladesh, given that fully one-third of the country is less than 3 meters above sea level. As of 2009, 105.6 million hectares of Bangladesh were suffering from saltwater infiltration, an area that represents over a quarter of the nation's land area. Saltwater intrusion not only damages crop yields, it contributes to high maternal blood pressure and pre-eclampsia, contributing to Bangladesh's high infant mortality rate. Many of the 400,000 low-income migrants drawn to the Bangladeshi megacity of Dhaka each year are in essence climate refugees, forced to flee by the mounting environmental problems of Bangladesh's shrinking and salinating coastline. By 2013, McDonnell claims, a projected 13.3 million Bangladeshis may be displaced by global warming.

This is not to say that Bangladesh is in any way helpless in the face of climate change. As Manoj Roy, Joseph Hanlon, and David Hulme (2016) point out, Bangladesh has overcome serious challenges in the past. At the time of its independence, some dismissed Bangladesh as "a famine-ridden 'basket case', but now it feeds itself," they note, thanks to widespread use of irrigation and locally developed strains of high yield rice. As we will see in Chapter 6, while Bangladesh owes its very origins in part to a devastating and deadly 1970 tropical cyclone, the death toll from such storms has fallen by 98% in the twenty-first century due to early warning systems and cyclone shelters. What is more, Bangladesh's farmers have also developed strategies of capturing river sediment to raise field levels, potentially allowing Bangladesh's coastal *char* islands to keep pace with global warming's rising seas. Failing that, coastal lands submerged by climate change could be transformed into shrimp and fish farms. Bangladesh therefore "confronts climate change from a basis of knowledge and experience." Still, Roy et al. worry that since climate change is being driven primarily by the greenhouse gas emissions of the industrialized western world, "the future of Bangladesh is partly in the hands of people far away, over whom it has no control and only limited influence."

In recent years, however, the part of the world that has become most closely associated with climate change is Syria, thanks in large part to an influential article by Kelley et al. in 2015 that linked the outbreak of the 2011 Syrian civil war to Anthropocene-era global warming. This is not to say that climate change acted alone. As these authors point out, a constellation of factors ensured that Syria was in a state of near crisis even before its time of troubles began in 2011. At the time of the famine, Syria was straining to support a massive influx of up to 1.5 refugees who had been displaced in 2003–2007 by an ongoing civil conflict in neighboring Iraq. Poorly thought-out agricultural policies played a role as well. The Syrian government had heavily subsidized irrigation projects and groundwater pumping to increase yields and generate more rural support for the regime, but since extraction of groundwater rose to 160% of the nation's renewable water supply, wells were drying up throughout Syria as the water table plummeted. Worst of all, Syria had suffered below-average rainfall in 7 of the 11 years that preceded the main 2007–2010 drought, and the rise of Syria's temperatures during this period had exceeded the global average, threatening crops like cereals that are dependent on soil moisture.

All of these problems came to a head during the severe 2007–2010 drought, the region's "worst 3-year drought in the instrumental record," which contributed to "massive agricultural failures and livestock mortality" Kelley et al. (2015). During these years many as 1.5 million farmers fled the desiccating countryside to the periphery of Syria's urban centers, Kelley et al. claim, and their desperate poverty helped to ignite the flames of civil war when the Arab Spring protests of 2010–2011 spread to Syria. Nonetheless, while the goals of the Arab Spring protesters in Syria might have been economic and political in nature, these authors contend that the triggering factor was climate change. Based on their model simulations, "a drought of the severity and duration of the recent Syrian drought … has become more than twice as likely as a consequence of human interference in the climate system," a finding that (they say) has relevance both for Syria's recent past, and for the world's near future.

Kelley et al.'s linkage between climate change and the Syrian civil war has been seconded by a number of other academics, including Peter H. Gleick, whose 2014 article on the Syrian famine and civil war made similar claims but did not generate the same headlines. However, some social scientists are not convinced. Political scientist Jan Selby (2019; 2020), for example, has discounted the role of climate change in the Syrian civil war, blaming it instead on long-term structural problems in Syria's economy, in particular, a failed oil rent-led model of national development. While not dismissing global warming's "physical reality and looming environmental threat" to the modern world, Selby notes that climate change is also "an object of discourse, debate and rhetoric, a potent meta-narrative that can be invoked for explanation, legitimation, blame avoidance and enrichment. Blaming the climate has long been, like blaming nature, a powerful rhetorical strategy," which has already been "regularly invoked to questionable ends across the Middle East and North Africa." Invoking climate change to explain a catastrophe, therefore, is the modern equivalent of claiming it is an "act of God," and Selby believes it serves the same purpose: to distract attention from the culpability of elite groups in an unfolding disaster.

In any case, the impact of climate change on history cannot be appreciated by looking at one-off events like the Syrian civil war. Climate change can cause extreme events, but so can random chance, so it is impossible to distinguish between a weather event triggered by climate change from an event that is extreme but falls within the normal range of possible weather. Rather, the full impact of climate change on human society will have to be calculated in aggregate, by seeing to what degree extreme climate events become more frequent as global temperatures rise. Even if global warming does cause events on the scale of Syria to become increasingly common – and I suspect they will – that does not absolve historians from the responsibility from treating each disaster as a discrete event. As we saw earlier in this chapter, apparently global climate events like the Little Ice Age can lead to quite different regional outcomes due to their complex interaction with local circumstances and vulnerabilities.

There is every reason to think that the same thing will be true about global warming as well, though perhaps with a modern twist. As Karen O'Brien and Robin Leichenko (2000) point out, contemporary climate change is occurring during an age of globalization and increasing worldwide economic integration. Economic globalization, however, has created clear winners and losers, enriching some countries (such as the EU, North America, and East Asia) while marginalizing others (like Sub-Saharan Africa and South Asia). The populations that will be worst impacted by the Anthropocene, therefore, will most likely be the two-time losers, groups are suffering simultaneously from both severe climate change *and* from economic marginalization and exploitation. O'Brien and Leichenko have dubbed this phenomenon "double exposure," and they argue that this concept might help researchers identify groups that are most vulnerable to climate change, such as coastal agriculturalists and impoverished urban slum-dwellers living in floodplains or perched upon steep slopes. If these authors are correct, modern-day climate change is likely to further worsen an ongoing global problem: the yawning gap between rich and poor, which during the modern era has been increasing both between the world's nations and within them.

Historians and other social scientists, therefore, are becoming increasingly aware that climate change has been a force to be reckoned with in world history. As the study of paleoclimates matures, evidence is mounting that both short-term climate impacts (especially the ENSO and volcanism) and longer-term fluctuations in radiative forcing have left their mark on past societies. However, this by no means invalidates the contribution of historians to the study of climate change. While historians are not experts in tree rings or oxygen isotopes in cave formations, we do have a special insight into the "deep history" of societies, such as socioeconomic structures, political fault lines, cultural frameworks, and short-term contingency and *conjuncture* – precisely the factors that determine a given society's vulnerability to climate change. As we have seen, historians can enrich the work of climate scientists by providing explanations for why Ming China experienced the Little Ice Age so differently from Tokugawa Japan, or why monsoon failures were so more deadly in the age of the British Raj than they were in the earlier Mughal era. The best way for historians to exorcise the ghost of Huntington, therefore, is to engage with climate science while continuing to do what we do best: providing a thick and nuanced understanding of the human context in which specific climate events unfold.

Further Reading

Adams, Nancy K. *et al.* "The Physical Volcanology of the 1600 Eruption of Huaynaputina, Southern Peru." *Bulletin of Volcanology*, vol. 62 (2001), pp. 493–518.

Crosby, Alfred W. *Children of the Sun: A History of Humanity's Unappeasable Appetite for Energy*. New York: W. W. Norton, 2006.

Dessler, Andrew E. and Edward A. Parson. *The Science and Politics of Global Climate Change*. New York: Cambridge University Press, 2006.

Dillon, Michael. *China: A Modern History*. New York: I. B. Tauris, 2012.

Dunning, Chester. *Russia's First Civil War: The Time of Troubles and the Founding of the Romanov Dynasty*. University Park, PA: Pennsylvania State University Press, 2001.

Fagan, Brian. *Floods, Famines, and Emperors: El Niño and the Fate of Civilizations*. New York: Basic Books, 1999.

Fan, Ka-wai. "Climatic Change and Dynastic Cycles in Chinese History: A Review Essay." *Climatic Change*, vol. 101 (2010), pp. 565–573.

Flannery, Tim. *The Weather Makers: How Man Is Changing the Climate and What It Means for Life on Earth*. New York: Atlantic Monthly, 2005.

Fleming, James Rodger. *Historical Perspectives on Climate Change*. New York: Oxford University Press, 1998.

Headrick, Daniel. "Global Warming, the Ruddiman Thesis, and the Little Ice Age." *Journal of World History*, vol. 26, no. 1 (2015), pp. 157–160.

Helama, Samuli, Phil D. Jones, and Keith R. Briffa. "Dark Ages Cold Period: A Literature Review and Directions for Future Research." *The Holocene*, vol. 27, no. 10 (2017), pp. 1600–1606.

Huntington, Ellsworth. *Civilization and Climate*. New Haven, CT: Yale University Press, 1924.

Issar, Arie S. "Climate Change and the History of the Middle East." *American Scientist*, vol. 83, no. 4 (1995), pp. 350–355.

Keys, David. *Catastrophe: An Investigation into the Origins of the Modern World*. New York: Random House, 1999.

Kolbert, Elizabeth. *Field Notes from a Catastrophe: Man, Nature, and Climate Change*. New York: Bloomsbury, 2006.

Ladurie, Le Roy E. *Times of Feast, Times of Famine: A History of Climate Since the Year 1000*. London: Allen & Unwin, 1971.

Mantua, Nathan J. and Steven R. Hare. "The Pacific Decadal Oscillation." *Journal of Oceanography*, vol. 58 (2002), pp. 35–44.

Marathe, Shamal and Ashok Karumuri. "The El Niño Modoki." In Swadhin Kumar Behera (Ed.), *Tropical and Extratropical Air-Sea Interactions: Modes of Climate Variations*. Amsterdam: Elsevier, 2019, pp. 93–114.

Mikhail, Alan. "Climate and the Chronology of Iranian History." *Iranian Studies*, vol. 49, no. 6 (2016), pp. 963–972.

Murphy, Robert Cushman. *Oceanic Birds of South America*. Vol 1. New York: Macmillan, 1936.

Nash, Madeleine J. *El Niño: Unlocking the Secrets of the Master Weather-Maker*. New York: Warner Books, 2003.

Oppenheimer, Clive. *Eruptions that Shook the World*. New York: Cambridge University Press, 2011.

Peregrine, Peter N. "Climate and Social Change at the Start of the Late Antique Little Ice Age." *The Holocene*, vol. 30, no. 11 (2020), pp. 1643–1648.

Perkins, Sid. "Disaster Goes Global: The Eruption of 1600." *Science News*, August 30, 2008, pp. 17–20.

Reid, Anthony. "The Seventeenth-Century Crisis in Southeast Asia." *Modern Asian Studies*, vol. 24, no. 4 (1990), pp. 639–659.

Reilly, Benjamin. *Disaster and Human History: Case Studies in Nature, Society, and Catastrophe*. 2nd ed. Boone, NC: McFarland, 2022.

Rosen, William. *The Third Horseman: Climate Change and the Great Famine of the 14th Century*. New York: Viking, 2014.

Selby, Jan. Omar S. Dahi, Christiane Fröhlich, and Mark Hulme. "Climate Change and the Syrian Civil War Revisited." *Political Geography*, vol. 60 (2017), pp. 232–244.

Sen, Amartya. *Poverty and Famines: An Essay on Entitlement and Deprivation*. New York: Oxford University Press, 1982.

Sibley, David J. *The Boxer Rebellion and the Great Game in China*. New York: Hill and Wang, 2012.

Wang, Xunming *et al.* "Climate, Desertification, and the Rise and Collapse of China's Historical Dynasties." *Human Ecology*, vol. 31, no. 1 (2010), pp. 157–172.

Works Cited

Atwell, William S. "Some Observations on the 'Seventeenth-Century Crisis' in China and Japan." *Asian Studies*, vol. 45, no. 2 (1986), pp. 223–244.

Behringer, Wolfgang. "Climatic Change and Witch-Hunting: The Impact of the Little Ice Age on Mentalities." *Climatic Change*, vol. 43 (1999), pp. 335–351.

Benson, Larry V. *et al.* "Possible Impacts of Early 11th-, Middle-12th-, and Late-13th-Century Droughts on Western Native Americans and the Mississippian Cahokians." *Quaternary Science Reviews*, vol. 26 (2007), pp. 336–350.

Brooke, John L. *Climate Change and the Course of Global History: A Rough Journey*. New York: Cambridge University Press, 2014.

Bulliet, Richard. *Cotton, Climate, and Camels in Early Islamic Iran: A Moment in World History*. New York: Columbia University Press, 2009.

Büntgen, Ulf *et al.* "Cooling and Societal Change During the Late Antique Ice Age from 536 to 660 A.D." *Nature Geoscience*, vol. 9 (2016), pp. 231–236.

Butzer, Karl W. "Collapse, Environment, and Society." *PNAS*, vol. 109, no. 10 (2012), pp. 3632–3639.

Cohen, Paul A. *History in Three Keys: The Boxers as Event, Experience, and Myth*. New York: Columbia University Press, 1997.

Cook, Edward R., Richard Seager, Mark A. Cane, and David W. Stahle. "North American Drought: Reconstructions, Causes, and Consequences." *Earth Science Reviews*, vol. 41 (2007), pp. 93–134.

Curtis, Valerie, Mícheál de Barra, and Robert Augner. "Disgust as an Adaptive System for Disease Avoidance Behavior." *Philosophical Transactions of the Royal Society of London, B*, vol. 366 (2011), pp. 389–401.

Davis, Mike. *Late Victorian Holocausts: El Niño Famines and the Making of the Third World*. New York: Verso, 2001.

De Waal, Alex. *Famine Crimes: Politics and the Disaster Relief Industry in Africa*. Indiana, IN: Indiana University Press, 1997.

Dinero, Stephen C. *Living on Thin Ice: The Gwich'in Natives of Alaska*. Brooklyn, NY: Berghahn Books, 2016.

Drake, B. Lee. "Changes in North Atlantic Oscillation Drove Population Migrations and the Collapse of the Western Roman Empire." *Scientific Reports*, vol. 7 (2017). doi:10.1038/241598-017-01289-z.

Drèze, Jean. "Famine Prevention in India." In Jean Drèze and Amartya Sen (Eds.), *The Political Economy of Hunger*. Oxford: Clarendon Press, 1990.

Dunning, Chester. "Crisis, Conjuncture, and the Causes of the Time of Troubles." *Harvard Ukrainian Studies*, vol. 19 (1995), pp. 97–119.

Ellenblum, Ronnie. *The Collapse of the Eastern Mediterranean: Climate Change and the Decline of the East*. Cambridge: Cambridge University Press, 2012, pp. 950–1072.

Fagan, Brian. *The Little Ice Age: How Climate Made History, 1300–1850*. New York: Basic Books, 2000.

Fagan, Brian. *The Great Warming: Climate Change and the Rise and Fall of Civilizations*. New York: Bloomsbury, 2008.
Fleming, James Rodger. "Climate, Change, History." *Environment and History*, vol. 20, no. 4 (2014), pp. 577–586.
Gleick, Peter H. "Water, Drought, Climate Change, and Conflict in Syria." *Weather, Climate, and Society*, vol. 6, no. 3 (2014), pp. 331–340.
Gowdy, John. "Our Hunter-Gatherer Future: Climate Change, Agriculture and Uncivilization." *Futures*, vol. 115 (2020). doi:10.1016/j.futures.2019.102488.
Grove, Richard H. "Revolutionary Weather: The Climatic and Economic Crisis of 1788–1795 and the Discovery of El Niño." In Robert Costanza and Lisa J. Graumlich (Eds.), *Sustainability or Collapse? An Integrated History and Future of People on Earth*. Cambridge, MA: MIT Press, 2007, pp. 151–167.
Grove, Richard H. and George Adamson. *El Niño in World History*. London: Palgrave Macmillan, 2018.
Hsiang, Soloman M., Kyle C. Meng, and Mark A. Cane. "Civil Conflicts Are Associated with the Global Climate." *Nature*, vol. 476 (2011), pp. 438–441.
Huntington, Ellsworth. "Changes of Climate and History." *American Historical Review*, vol. 18, no. 2 (1913), pp. 213–232.
Issar, Arie S. and Mattanyah Zohar. *Climate Change: Environment and History of the Near East*. 2nd ed. New York: Springer, 2007.
Jacobsen, Thorkild and Robert M. Adams. " Salt and Silt in Ancient Mesopotamia Agriculture." *Science*, vol. 128, no. 3334 (1958), pp. 1251–1258.
Kelley, Colin P. *et al.* "Climate Change in the Fertile Crescent and Implications of the Recent Syrian Drought." *Proceedings of the National Academy of Sciences*, vol. 112, no. 11 (2015), pp. 3241–3245.
Kemp, Luke *et al.* "Climate Endgame: Exploring Catastrophic Climate Change Scenarios." *PNAS*, vol. 119, no. 34 (2022). doi:10.1073/pnas.2108146119.
Lamb, H. H. *Climate, History and the Modern World*. London: Routledge, 1982.
Lee, Jeffrey A. and Thomas E. Gill. "Multiple Causes of Wind Erosion in the Dust Bowl." *Aeolian Research*, vol. 19 (2015), pp. 15–36.
Lefebvre, Georges. *The Great Fear of 1789: Rural Panic in Revolutionary France*. Princeton, NJ: Princeton University Press, 1973.
Linden, Eugene. *The Winds of Change: Climate, Weather, and the Destruction of Civilizations*. New York: Simon & Schuster, 2006.
Matossian, Mary Kilbourne. *Poisons of the Past: Molds, Epidemics, and History*. New Haven, CT: Yale University Press, 1989.
McCormick, Michael *et al.* "Climate Change during and after the Roman Empire: Reconstructing the Past from Scientific and Historical Evidence." *Journal of Interdisciplinary History*, vol. 43, no. 2 (2012), pp. 169–220.
McDonnell, Tim. "Climate Change Creates a New Migration Crisis for Bangladesh." *National Geographic*, January 24, 2019.
McMichael, Anthony J. *Climate Change and the Health of Civilizations*. New York: Oxford University Press, 2017.
Miller, Gifford H. *et al.* "Abrupt Onset of the Little Ice Age Triggered by Volcanism and Sustained by Sea-Ice/Ocean Feedbacks." *Geophysical Research Letters*, vol. 39 (2012). doi:10.1029/2011GL050168.
O'Brien, Karen L. and Robin M. Leichenko. "Double Exposure: Assessing the Impacts of Climate Change within the Context of Economic Globalization." *Global Environmental Change*, vol. 10 (2000), pp. 221–232.

Parker, Geoffrey. *Global Crisis: War, Climate Change and Catastrophe in the Seventeenth Century*. London: Yale University Press, 2013.

Pfister, Christian. "The Vulnerability of Past Societies to Climatic Variation: A New Focus for Historical Climatology in the Twenty-First Century." *Climatic Change*, vol. 100 (2010), pp. 25–31. doi:10.1007/s10584-010-9829-2.

Preston, Diana. *The Boxer Rebellion*. New York: Walker & Co., 2000.

Rabbani, Golam, Sirazoom Munira, and Samia Saif. "Coastal Community Adaption to Climate Change-Induced Salinity Intrusion in Bangladesh." In Surenda N. Kulshreshtha (Ed.), *Agricultural Economics*. New Delhi: Books on Demand, 2019. doi:10.5772/intechopen.78437.

Roy, Manoj, Joseph Hanlon, and David Hulme. *Bangladesh Confronts Climate Change: Keeping Our Heads Above Water*. New York: Anthem Press, 2016.

Ruddiman, William. *Plows, Plagues, and Petroleum: How Humans Took Control of Climate*. Princeton, NJ: Princeton University Press, 2005.

Selby, Jan. "Climate Change and the Syrian Civil War, Part II: the Jazira's Agrarian Crisis." *Geoforum*, vol. 101 (2019), pp. 260–274.

Selby, Jan. "On Blaming Climate Change for the Syrian Civil War." *Middle East Research and Information Project*, vol. 296 (2020). Available at: https://merip.org/2020/10/on-blaming-climate-change-for-the-syrian-civil-war-296/

Simonow, Joanna. *Ending Famine in India: A Transnational History of Food Aid and Development*. Leiden: Leiden University Press, 2023.

Sœther, Bjørnar and Karen O'Brien. "'The Hottest Summer Ever!' Exploring Vulnerability to Climate Change among Grain Producers in Eastern Norway." In Greg Bankoff and Dorothea Hilhorst (Eds.), *Why Vulnerability Still Matters: The Politics of Disaster Risk Creation*. New York: Routledge, 2022, pp. 222–237.

Trevor-Roper, H. R. "The General Crisis of the 17th Century." *Past and Present*, no. 16 (1959), pp. 31–64.

Vavrus, Stephen J. *et al.* "Did Agriculture Beget Agriculture During the Past Several Millennia?" *The Holocene*, vol. 32, no. 7 (2022), pp. 680–689.

Waldinger, Maria. "Let Them Eat Cake: Drought in 1788 and Political Outcomes in the French Revolution." Ifo Institute, Munich, Working Paper, July 19, 2021.

Watson, Andrew M. *Agricultural Innovation in the Early Islamic World*. New York: Cambridge University Press, 1983.

White, Sam. *The Climate of Rebellion in the Early Modern Ottoman Empire*. New York: Cambridge University Press, 2011.

Worster, Donald. *Dust Bowl: The Southern Plains in the 1930s*. New York: Oxford University Press, 1979.

Zhang, David D. *et al.* "Climatic Change, Wars and Dynastic Cycles in China over the Last Millennium." *Climatic Change*, vol. 76 (2006), pp. 459–477. doi:10.1007/s10584-005-9024-z.

Zhang, David D. *et al.* "Climate Change and War Frequency in Eastern China over the Last Millennium." *Human Ecology*, vol. 35, no. 4 (2007), pp. 403–414.

Zhang, David D. *et al.* "The Causality Analysis of Climate Change and Large-Scale Human Crisis." *PNAS*, vol. 108, no. 42 (2011), pp. 17296–17301.

6 Tropical Cyclones

It's late evening on Labor Day, 1935. As the winds howl around you on Florida's Lower Matecumbe Key, you think back ruefully upon the chain of events that put you here in the path of the storm. Eighteen years before, you had been drafted and shipped off to France to fight the Germans in the trenches of the Great War. Although the traumas of that terrifying time stalk your dreams and complicated your return to civilian life, at least they qualified you for the "bonus" that Congress promised to pay Great War veterans in 1945. But once the Great Depression struck in 1929, 15 years became too long to wait. Unemployed and hungry, you needed that bonus sooner rather than later, so in 1932 you joined your veteran comrades in a "Bonus Camp" on the outskirts of Washington and staged demonstrations against the tight-fisted Hoover administration and its refusal to release the money that was your due.

As the savage, sand-blasting winds dismantle the labor camp around you, you are reminded of what Hoover did to your Bonus Camp back in 1932. You and your fellows were chased from it by advancing tanks and clouds of tear gas, and the army set fire to what you left behind. Thankfully, that bastard Hoover lost the next election to Roosevelt and the Democrats. Not that FDR was much better; when you and many other "Bonus Marchers" returned to Washington in 1933, he still refused to give you your bonus. But he did enroll you and your fellows in the newly created Civilian Conservation Corps, which promised the Bonus Marchers paid employment in government-funded public works schemes.

To your surprise, the winds suddenly die down to balmy breezes, which are reminiscent of the breezes that greeted you when you first came to Florida in the employ of the CCC. In order to get you and your fellow Bonus Marchers as far from Washington as possible, FDR had banished you to the remote Florida Keys, where the CCC was constructing a new highway down the island chain to Key West. The work camps that the CCC built in the Florida Keys were nowhere close to paradise: there was not enough fresh water, far too many mosquitoes, and no entertainment but booze from the canteen and the carnal

DOI: 10.4324/9781003436805-7

pleasures of Key West at the end of the rail line. The work was hard and monotonous beneath an unrelenting subtropical sun. But at least you had a job, and a roof over your head. Or you did until earlier that evening, when mounting gales of the incoming hurricane had ripped it away and sent you running into the mangroves for shelter.

Now that your train of memory has brought you back to the present, you settle down in the thicket and listen for the rescue train, which the CCC had promised to send if the camps were threatened by a hurricane. But all you hear is the renewed roar of the wind, which is returning with even greater strength than before. Water begins to rise around you, reaching your ankles, then your kneecaps, now higher still. And what is that sound? You hear a dreadful, gathering growl, like the thunder of an artillery barrage, approaching in the darkness on the Atlantic side of the island. Just as this roar reaches its crescendo, you see it in the distance with every lightning flash: an impossibly large incoming wave, rising higher than the brothels and saloons of Key West, heading straight toward Lower Matecumbe Key.

With terrible suddenness, the wave crashes into you and pries you loose from the mangroves. As the floodwaters sweep you toward Florida Bay, you wonder, am I going to die? And if so, what killed me – the Great Depression, the Bonus March, the Roosevelt administration, or the Labor Day Hurricane?

Cyclone Science

The Labor Day Hurricane of 1935 was a particularly strong exemplar of a more general phenomenon: cyclonic weather systems, which, broadly defined, are any meteorological system in which a large air mass rotates around a low-pressure core. Technically, tornadoes and waterspouts are cyclones, as are "extratropical" cyclones, such as the powerful "nor'easter" winter storms of New England and Atlantic Canada. This chapter, however, will focus on one specific and particularly important category of cyclones: tropical cyclones.

Although they are known by regional names, such as hurricanes in the Atlantic, typhoons in the northwestern Pacific, and cyclones in the South Pacific and the Indian Ocean, all these terms refer to the same phenomenon: powerful storms born from the interaction between convection and the Coriolis effect. Cyclones can only form over the ocean, and only when the water temperature has reached about 26–27°C. As a result, cyclones are highly seasonal and regional: typically, only tropical seas close to the equator reach this temperature threshold, and only in the Northern Hemisphere's summer and early fall. For their genesis, cyclones also require a seed in the form of a low-pressure system of some kind. The Labor Day Cyclone, for instance, probably started life as a low-pressure cell generated by the heat differential between the hot Sahara of northern

Africa and the cooler forests of coastal West Africa. This cell was then blown into the Atlantic Ocean by the prevailing westward-flowing trade winds of the equatorial region.

Whether or not one of these "seed" low pressure cells generates a cyclone depends not only on the water temperature of the ocean below, but also the distance from the equator. Since hurricanes are essentially localized convection currents, they require the energy of warm water to strengthen, so low pressure cells that blow too far north over colder seas generally lack the fuel to transform into cyclones. That being said, a growing body of literature reviewed by Studholme et al. (2022) does suggest that global warming will push the limits of cyclone formation northwards and southwards from the equator in the coming decades. At the same time, since hurricane formation depends on the spin given to large weather systems by the Coriolis effect, low-pressure systems that form near the equator, where Coriolis forces are weak, generally cannot strengthen into cyclones. As a result, the butter zone for hurricane formation lies between 5° and 20° latitude in the Northern and Southern Hemispheres.

The process of cyclone formation begins when low pressure at the heart of the seed cell encourages the evaporation of warm ocean water, creating a rising column of moist air. The relative vacuum thus created draws in surrounding air masses, but instead of filling the low pressure of the cell, they are deflected to the left (right in the Southern Hemisphere) by the Coriolis effect, giving the strengthening storm a counter-clockwise (clockwise) spin. There are many factors that can interrupt cyclone formation at this point of the process. High-level wind shear can effectively behead the storm and short-circuit the convection cycle. Cyclones can also lose steam if they wander over a cold patch of seas or pass over dry land. Absent of such interruptions, the cyclone will steadily strengthen, becoming a tropical depression, then a tropical storm, and finally a tropical cyclone, each of which marks an exponential increase in the overall power of the storm.

Once fully mature, cyclones take on a characteristic spiral shape, which can extend over 800 km from the storm center or more in the case of large Pacific typhoons. Of course, the heart of the cyclone is its eye, the bowl of low pressure around which the cyclone rotates. According to David Longshore (2000), the eye of a mature cyclone serves as an "efficient airshaft, an umbilical cord through which the storm will grow from the inside out." Moisture-laden air spirals upwards around the rim of the eye, and when it reaches the colder upper atmosphere, the water vapor condenses back into liquid water droplets, which form the "rain band" cloud structures that make up most of the pinwheel body of a mature cyclone. The process of condensation also releases energy, helping to fuel the strength of the cyclone's winds. Nonetheless, when this eye is passing overhead, the storm may seem to suddenly dissipate, and those caught in the well-defined eye of a powerful storm might even see blue skies or stars overhead. But as we saw in the Labor Day Hurricane above, the respite is only temporary. While the eye itself is calm, the clouds of the eyewall that surround it generally harbor the cyclone's strongest winds. Worst of all, when the winds return after the passage of the eye, they blow in the opposite direction, placing new stresses on structures that were already weakened in the first half of the storm, and posing new peril to those lured outside by the eye's deceitful calm.

Once formed, cyclones tend to move in fairly predictable directions, depending on the seas that spawned them. In the Atlantic and Pacific Oceans, cyclones tend to wander westwards toward the American and East Asian land mass respectively, carried along by the prevailing westward-blowing trade winds of the equatorial region. Indian Ocean cyclones, in contrast, are more influenced by the movement of the Intertropical Convergence Zone (ITCZ); indeed, the monsoon season in the Indian Ocean largely overlaps with the cyclone season. In all seas, cyclone movement is also tweaked to the right (left in the Southern Hemisphere) by the Coriolis force, and this influence over cyclonic storm tracks becomes more pronounced the farther the cyclone moves from the equator (Figure 6.1). The path taken by cyclones is also strongly influenced by anticyclones, high-pressure cells like the Azores High of the Atlantic subtropics mentioned in Chapter 5. Depending on the strength and location of this high, Atlantic hurricanes can start recurving to the north anywhere between 60th and 90th west longitude, meaning that any given hurricane can arc back into the Atlantic and miss land entirely, scrape the east coast of the United States, wash ashore in the US Gulf Coast, or keep tracking westward along the Caribbean islands and eventually plow into Central America or Mexico. Similarly, the location and strength of the high-pressure subtropical ridge in the North Pacific help to determines whether a Pacific typhoon will track westward into the Philippines, South China,

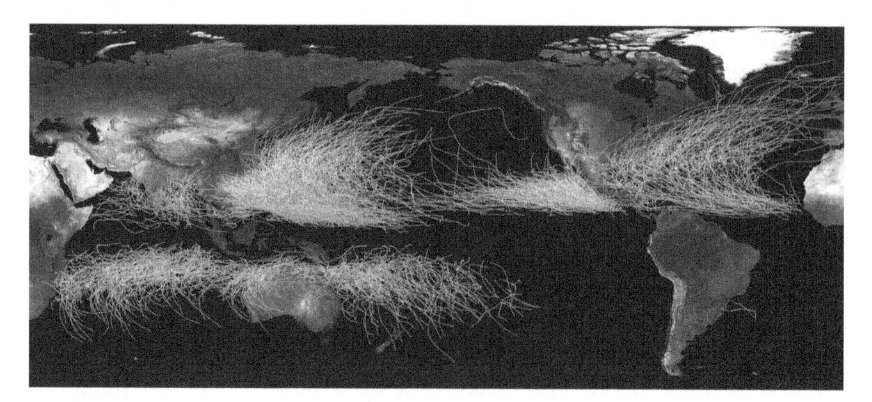

Figure 6.1 Tracks of tropical cyclones, 1985–2005
Source: Wikimedia Commons, created by Nilfanion.

Taiwan, and Indochina, or will recurve northwards, potentially striking the Eastern Philippines, Eastern China, Japan, and the Korean Peninsula.

When these cyclones do wash ashore, they pose a number of different hazards to human societies caught in their path. Cyclonic winds can cause extensive damage to homes and other structures, especially in the "dirty side" of a cyclone, the front right quadrant of the storm (front left in the Southern Hemisphere), where the rotational motion of the storm is moving in the same direction as the storm itself. In the case of a category-5 cyclone, such as the Labor Day Hurricane described above, wind speeds might reach 295 km/m. Observers to that storm remembered that the debris blown by the wind was as deadly as bomb shrapnel, and one survivor remembered seeing "bodies with tree stumps smashed through their chests – heads blown off – twisted arms and legs, torn off by flying timber that cut like big knives." Cyclones often lead to chain reactions of destruction, where the disintegration of one building creates flying debris that smashes into other structures, contributing to yet more building failures and further wind-blown debris.

The same winds also contribute to the most deadly hazard that cyclones pose: salt-water coastal flooding, or "storm surge." Because of the low atmospheric pressure at the center of a cyclone, the seawater below the storm bulges upwards to fill the relative vacuum. This bulge is stacked higher by the storm's prevailing winds, especially in the "dirty" front-right quadrant of the cyclone. Since the height of the storm surge correlates with the strength of the cyclone, the storm surge produced by a category-5 cyclone can strike with much the same force as a powerful tsunami, though generally over a smaller area. The Labor Day Hurricane's "tidal wave," for example, topped out at 7.5 meters above sea level, and it accounted for the vast majority of the storm's 422 deaths. As we will see below, even larger storm surges have been reported in the Indian Subcontinent, especially the Ganges Delta, where the death tolls from cyclonic storm surges are counted in the tens of thousands rather than hundreds.

A final cyclone hazard is torrential rainfall and disastrous freshwater flooding. In October of 1998, for example, Hurricane Mitch dropped almost a meter of rain as it passed over parts of Central America, leading to widespread flooding and hundreds of landslides. In Honduras, 7,000 people were killed by floods and mudslides, and the estimated damage was over $3.8 billion, a total equal to 70% of Honduras GDP at the time. In nearby Nicaragua, over 3,800 were killed, many of them during a lahar triggered when volcanic materials deposited by the Casita Volcano were loosened by Mitch's deluges. It should be noted that, in both Nicaragua and Honduras, Mitch's rains fell upon a landscape that was already vulnerable due to the region's political and socioeconomic problems. Both countries are among the poorest in the Western Hemisphere, and in Honduras, widespread pre-cyclone deforestation greatly worsened the impact of the floods and landslides. In Nicaragua, clean-up after the landslides was complicated by the fact that up to 75,000 live landmines left over from the U.S.-funded Contra insurgency of the 1980s had been displaced by the floodwaters! As for the Labor Day Hurricane, rainfall contributed little to the death toll in the Florida Keys, but this powerful storm did dump up to 410 millimeters of rain as it passed up the interior of the US east coast, contributing to substantial crop losses from Georgia all the way to Delaware.

Cyclones and Cultures of Disaster

Arguably, the part of the world with the earliest "culture of cyclones" is China. According to cyclone researchers Kin-sheun Louie and Kam-biu Liu (2004), the Chinese had coined the term "*jufeng*," meaning a "wind that comes in all four directions," by the fifth century CE. *Jufengs* were also defined as a "scary winds," and although it is tempting to interpret this as Chinese understanding of the dangers that cyclones pose, it might just be a pun: the Chinese word *ju*, meaning wind or storm, is a homophone for the Chinese word for "scary." In any case, Louie and Liu note that the Chinese seem to already have had a culture of disaster by the fifth century when it comes to cyclonic weather. In addition to understanding that cyclones winds traveled in a circular motion, contemporary chroniclers were already distinguishing between major and minor cyclones, and they claimed that the coming of a cyclone could be predicted by the behavior of "roosters and dogs," who would go silent in the three days before a cyclonic storm struck. The same chronicler also noted that the same storms are called "*heifing*" or "black winds" in other countries, suggesting that the Chinese were not the only Asian society that recognized cyclones as a recurrent hazard. It should be noted that the term "typhoon" seems to have appeared in the twelfth century, and while most dictionaries define it as an Indigenous Chinese word meaning "winds which long last," a minority view holds that the Chinese derived their word from the Arabic word "*tufan*," or "smoke."

Given the antiquity of China's culture of cyclone, it is not surprising that the first unequivocal record of a cyclone strike comes from the annals of Chinese history. According to a historical chronicle of the Tang dynasty, a *jufeng* struck the city of Mizhou (today Gaomi, in Shandong) in 816 CE and damaged the city wall with its storm surge. Just three years later, China would record a second cyclone strike. This time, the cyclone struck southern China, and it was immortalized by the Tang poet Han Yun:

> Encountering jufeng at Shashan
> thunder and lightning struck hard;
> I then arrived at Fuxu among strong tidal waves
> the scene was terrible on the shore;

although the two bluffs looked sturdy
wood and rock debris were flying around;
although Tunmen seemed high
It was submerged in the choppy water.

In addition to leaving its mark on Chinese poetry, the *jufeng* exerted a strong influence on Chinese settlement patterns, at least in the cyclone-prone south. According to Jin Tao et al. (2022), villages in southern China's Leizhou Peninsula tried to mitigate cyclone hazards using multiple techniques, such as planting windbreak forests, narrowing the side of their houses that faced the prevailing cyclonic winds, and building on a slight slope to facilitate drainage. Most importantly, the Chinese of Leizhou Peninsula traditionally build their homes in a "dense alley" layout, creating what local Chinese called a "comb village," featuring many parallel alleys that, from above, resemble the teeth of a comb. In such villages, houses are built close together in a grid pattern, so that the houses on the outside of the grid shield the interior houses from cyclonic winds, and these houses were often interconnected for mutual reinforcement. The large number of alleyways traversing the village, in turn, allows for quick drainage of the village after being soaked by cyclonic rainfall. As for the roofs, they are typically built to a slope of about 37° in these "dense alley" villages, which Tao et al. believe reflects a compromise between providing resistance to the wind (best accomplished by a low roof) and waterproofing the house (better accomplished with a peaked roof). In addition, Leizhou Peninsula houses tend to have heavy tiled roofs and few protruding eaves to reduce the chance of the roof being ripped away by cyclonic winds.

Amerindian groups in the Caribbean basin, who also had a long history of exposure to cyclonic weather, developed their own cultures of disaster surrounding North Atlantic cyclonic storms. The name "hurricane," by which these storms are now known in the Atlantic Ocean, probably derives from the Mayan god "Hurakán," the lord of wind, storm, and fire, who once in myth summoned a great flood to wipe out mankind. By the time European settlers arrived in the Americas, the Taíno people of the Caribbean seem to have assimilated Hurakán into their own culture as the *jurucán*, powerful wind spirits summoned by the goddess Guabancex, deity of chaos and disorder. Interestingly, Guabancex was typically depicted in iconography as the face of an angry woman surrounded by two arms extended in a spiral pattern, which as Stuart Schwartz (2015) points out forms an image strikingly similar to the spiral used today to symbolize a cyclonic storm on a weather map or a hurricane evacuation route sign. This might indicate that the Taíno understood the cyclical nature of such storms, though in the absence of a written record it is impossible to be sure.

According to archeologist Jago Cooper (2012), pre-Columbian Taíno groups developed a suite of strategies to mitigate Guabancex's fury. Taíno farmers tended to plant root crops like yucca rather than standing crops like maize, since the latter was much more vulnerable to cyclonic winds. When building houses, the Taíno began with a double ring of thick hardwood poles driven up to 1.7 meters into the ground and then added a "lighter-weight superstructure including a slender rafter, stringer, and thatched roof." While this superstructure was not hurricane-proof, the foundations themselves could withstand the storm, allowing for the house to be "rebuilt within days using locally available and easily sourced materials." Taíno villages also tended to be built in sheltered locations, such as the leeward side of hills, where they would have been spared the brunt of a westward-traveling hurricane's winds. They also avoided lowland areas with poor drainage and coastal cites exposed to the ocean, opting instead for locations shielded behind a protective layer of mangroves. In addition, nearly all settlement sites were located near caves, which served as hurricane shelters. As Cooper points out, most of these Taíno adaptations were forgotten during the immediate post-Columbian period. Rather, the colonizing Spanish founded their settlements in economically rewarding but environmentally risky locations such as river floodplains.

Hurricanes and History in the Sixteenth–Eighteenth-Century Caribbean

In contrast to the Taíno, the Europeans who colonized the Caribbean basin from the sixteenth century onward had almost no experience with cyclonic weather and therefore no culture of hurricane mitigation. As a result, when Spanish colonists realized that the "islands and coasts" of the Caribbean suffered from the "worst storms of all the world's seas," they initially blamed the savageness of the weather on the savagery of the region's Indigenous inhabitants. Chronicler of the Spanish East Indies Gonzalo Fernández de Oviedo, for example, suggested that the terrible storms of the region were a sign that the Americas were under the influence of demonic forces, but noted that, in areas where the Eucharist ceremony had been celebrated by Spanish Christians, the "hurricanes and great storms were no longer as frequent or as dangerous as they had been." Schwartz (2015) notes that the Spanish also used holy relics as talismans against American hurricanes. Such practices represent an Americanization of the widespread European Christian tradition of spiritual prophylaxis against disaster that we have already discussed in previous chapters, for example, in the construction of holy chapels in landslide-prone terrain. In general, therefore, the Spanish regarded hurricanes through the lens of theology rather than natural science, seeing them primarily as manifestations of divine wrath. The fact that Indigenous Amerindians seemed to be able to predict the coming of such storms was therefore interpreted, not as a sign of Indigenous ecological knowledge, but as further proof that Amerindian shamans were in league with the devil.

Over time, however, Spanish colonists moved from religious interpretation of cyclonic weather to a more practical understanding of these events grounded in Indigenous creole hurricane experience. The "wrath of God" explanation of hurricanes became harder to sustain when colonists realized that hurricanes were restricted to certain seasons of the year, while human sinfulness presumably was not. Perhaps because hurricanes seemed to be such a natural and predictable part of the Caribbean landscape, Schwartz notes, that they did not inspire the same "culture of fear and eschatological anguish" that was provoked by large earthquakes in the colonial Americas. Major earthquakes like the Santiago de Chile quake of 1647 and the Lima earthquake of 1746 were followed by memorials and "apocalyptic sermons," but Schwartz notes that such religious soul-searching was largely absent after hurricanes in the Spanish Caribbean. It should be noted that the contrast that Schwartz draws between the existential terror provoked by the earthquakes and the more pragmatic reaction of colonists to cyclones perfectly matches the findings of Jeanet Sinding Bentzen's (2019) article on the comparative psychological impact of different disasters. As we saw in Chapter 2, Bentzen has argued that unpredictable disasters like earthquakes are far more likely to inspire an uptick of religiosity than more predictable disasters such as cyclonic storms, and the historical data Schwartz brings to bear seems to support Bentzen's contention.

West Indian hurricanes may not have inspired strong religious reactions in Spanish and other European creoles, but they did require colonists to keep one eye on the natural world and the other on the calendar. Cyclonic weather, many Spanish creoles believed, could be predicted by watching for unusual animal behaviors. When frigate birds fly inland, colonists told each other, it was a sign of incoming bad weather, and in Puerto Rico it was said that "when crickets, cicadas, toads and frogs disappear, hurricane for sure." More generally, European vessels in the Caribbean tried to adhere to a rigid nautical timetable, leaving the West Indies in June or July, before the start of the hurricane season but late enough to catch favorable winds in the North Atlantic. As Adam Sobel (2014) points out, "the regularity of the schedule made the ships vulnerable to pirates and enemy fleets, but the storms were a worse danger." To further mitigate the risks posed by hurricanes, creole planters insured their cargos and divided them into smaller shipments that were parceled out into multiple vessels in an attempt to prevent a total loss of a season's harvest to shipwreck or piracy.

More generally, experience with the hurricane would leave its mark on Spanish settlement patterns and architectural practices. While some initial Spanish settlements were founded in vulnerable locations, such as riverine flood plains and exposed coasts, the Spanish would eventually relocate to more sheltered highland and inland locations. They also gravitated toward harbor sites that offered some protection from northwestward-tracking hurricanes, such as Havana on the northern coast of Cuba, which was far less exposed to these storms than the harbors in the island's south. In addition, Spanish creoles adapted their construction styles to the region's hurricane winds. Permanent buildings were designed to have a low profile and were constructed with stone masonry, reinforced by sturdy local hardwoods, and topped with a heavy tile roof. If sensibly situated, historian Sherry Johnson (2011) notes, such houses were "usually able to withstand the worst a storm could deliver." Spain's poorer subjects contented themselves with wattle and daub huts with a thatched roof. Houses of this sort were unlikely to survive even a moderate hurricane, but like the houses of the Taíno, they could be easily rebuilt with local materials in the hurricane's aftermath.

By the seventeenth century, Spain's dominance over the islands of the Caribbean was coming under threat from the French and the English, who steadily were carving out their own West Indian colonial empires on islands the Spanish had neglected or had been forced to surrender. Unfortunately for the French and English, when the Spanish left a territory, they took their hard-won culture of hurricanes with them, obliging the English and French interlopers to develop their own culture of hurricanes from scratch. Colonists in British Jamaica, for example, initially built their houses "in the English style," meaning multi-story structures made of brick or stone with gabled roofs. As Matthew Mulcahy (2006) points out, such structures frequently succumbed to hurricanes, while "the dwellings built by the Spanish before the English conquest of 1655 often withstood the worst storms." Over time, the English followed in the footsteps of the Spanish (and for that matter, the southern Chinese) and began to construct one- or two-story houses exclusively of brick and stone with flattened "hipped" roofs to lower their profile against hurricane winds. Houses like this were decidedly unfashionable – Mulcahy reports that one visitor to Barbados thought they were so devoid of ornamentation that they "conveyed the idea of barns" – but they were resistant to hurricane winds.

As further protection, some French and English plantation owners invested in "stormhouses" or "hurricane houses." Constructed of wood, masonry, or brick, such buildings were an artificial version of the caves used by Taíno as storm shelters in the Pre-Columbian era. Many of these shelters, like the huts of the Taíno, were constructed atop wooden posts sunk deep into the ground, though in this case the posts were probably intended to anchor the building in place rather than facilitate rapid post-storm reconstruction. In later years, the English even began to construct "storm towers" in areas vulnerable to cyclonic storm surge. These structures resembled squat brick silos and were elevated somewhat above the surrounding terrain, thus offering refuge from flash floods and storm surge as well as cyclonic winds.

In addition, the English and French, like the Spanish before them, recognized that the hurricanes were a sharply seasonal danger and organized their activities accordingly. According to an old Mariner's poem with origins in the British Caribbean,

> June too soon.
> July stand by.
> August look out you must.
> September remember.
> October all over.

Like the Spanish, English shippers insured their cargos and timed their voyages to avoid the late summer–early fall hurricane season. Ships inbound from Europe would try to arrive in the Caribbean by April, which not only was before the start of the hurricane season but also marked the end of the sugar-making season in the West Indies. During the dangerous summer months, captains took their vessels up navigable rivers for safety or even shipped out for South America, which had some storms but no hurricanes. Captains also knew which ports to avoid during the hurricane season, like Bridgetown in Barbados, and which harbors were relatively secure, such as St. Johns in Antigua, which was surrounded by a ring of sheltering hills.

As they developed their own "culture of hurricanes," the English, like the Taíno and Spanish before them, had to address the issue of causation: why did these terrible storms take place? Like the Spanish, the English initially looked at them through a religious lens, seeing in such storms the work of divine providence. However, while the Spanish tried to combat such spiritual threats with relics and the Holy Eucharist, the English put their faith in progress. As Mulcahy points out, seventeenth-century colonists blamed the wildness of the Caribbean weather on the savagery of the untamed landscape and believed that their "improvements" to the land would lead to a corresponding reduction to the risk of hurricanes. Eighteenth-century English scholars claimed that Caribbean forests were not only reservoirs of disease, they also blocked the purifying passage of the region's trade winds. As these wild forests were replaced by cultivated land, therefore, the air would become more wholesome and the number of hurricanes would dwindle. This, of course, was wishful thinking. In actual fact, British "improvements" to the landscape exacerbated the hurricane problems by removing natural windbreaks and rendering the landscape more vulnerable to erosion.

The Great Hurricanes of 1780: Slavery, Starvation, and the American Revolution

The hurricanes cultures of the English, Spanish, and other colonial powers would be put to an extreme test in the second half of the eighteenth century. Between 1750 and 1800, Johnson (2011) claims, the ENSO cycle veered into a strongly positive mode as the world emerged from the nadir of the Little Ice Age. Fully 31 of these 50 years witnessed "La Niña" episodes or events, 10 of which were severe. Since La Niña generally exacerbates cyclone formation in the North Atlantic, the Caribbean suffered repeated storms during these decades, especially during the extraordinarily active 1780 Atlantic hurricane season. At least eight hurricanes blew through the Caribbean or North Atlantic that year, directly killing about 28,000 people and condemning many thousands more to slower deaths by disease or starvation. In the process, Johnson and other authors argue, these cyclones played an important role in both the outbreak and the outcome of the War of American Independence.

The first great storm of the 1780 hurricane season struck "too soon" in June and caused considerable loss of life and property in Spanish Puerto Rico. The same storm also destroyed an English warship, which would become a recurring theme in the storms that followed. Months later, during the supposedly safe month of October, a "perfect tempest" smashed into Savanna-la-Mar in Jamaica, bringing with it a 6-meter storm surge that killed up to 3,000 people. The vessels in the harbor, including a number of British warships, were either destroyed at sea or driven inland, and some settlements along the Jamaican coast were obliterated almost entirely. In the words of one contemporary observer, the sea "broke suddenly in upon the town, and on its retreat swept every thing away with it, so as not to leave the smallest vestige of Man, Beast, or House behind." Their homes gone, some storm survivors resorted to washed-ashore shipwrecks for shelter against the elements in the storm's aftermath.

The most devastating cyclonic storm of the 1780 hurricane season, however, was the Great Hurricane of October 10. In Barbados, where it first touched shore, a British official recorded that:

> strongest buildings and the whole houses, most of which were stone and remarkable for their solidity, gave way to the fury of the wind and were torn up to their foundations … more than 6,000 perished and all of the inhabitants are entirely ruined.

Next to be struck was St. Vincent, which was inundated by a 6-meter storm surge. The Great Hurricane then recurved northwards into the Windward Islands, which at the time was a front line in an ongoing Anglo/French conflict set off by the American War of Independence. In St. Lucia, which the British had recently seized from the French, the Great Hurricane reportedly left only two houses standing on the whole island and nearly wiped out a whole British naval squadron, killing hundreds of British sailors in the process. The storm then changed its allegiance and slammed into the French island of Martinique, where it reportedly slew over 9,000 colonists and soldiers and sank 40 French vessels. The arcing path of this cyclone eventually took it back over the Atlantic Ocean, where it dissipated, but not before inflicting serious damage and loss of life in Spanish Puerto Rico and British Bermuda.

As Mulcahy (2006) points out, the population most afflicted by these repeated hurricanes was the region's enslaved Africans, who perished both during the storm and in its hungry aftermath. As is often the case in disasters, the short-term crisis triggered by the 1780 Great Hurricane had deeper roots. The Windward Islands traditionally imported much of their foodstuffs from the 13 colonies of the American mainland, but such shipments had ended in 1774 in the tense lead-up to the War of the American Revolution. By March of 1778, the English Governor William Burt of St. Kitts wrote to London complaining of spiraling food prices and a general "want of Provisions" in Antigua. Burt also claimed that, food shortages had contributed to the deaths of "above a Thousand Negros," and in Montserrat he claimed that "near twelve hundred" had died, including even "some whites." He therefore pleaded for the British to dispatch foodstuffs to the region, since "the Ground & Colonial Provisions" that were currently available "will not half Supply us," and Burt worried that "should there be a Hurricane God only knows what may be the Event." But the problems of food provisioning only got worse in June of 1778, after France and Spain declared war on Britain in support of the rebellious American colonists, turning the Windward Islands into a war zone and suppressing trade. Even before 1780, therefore, the British Caribbean was starving.

The Great Hurricane, Mulcahy claims, probably killed 2,000 slaves in Barbados on October 10, with most victims "crushed beneath falling buildings or drowned in storm surges." Many more would die when the pre-cyclone hunger escalated into a full-blown famine, just as Burt had predicted two years previously. In much of the Caribbean, slaves were expected to grow their own provisions, generally plantains and maize. However, both crops proved highly vulnerable to wind damage, leaving slaves with almost nothing to eat. As late as July 1781, a visitor to the "dismal ruins" of Bridgetown in Barbados discovered that "every article of produce is very scarce, particularly fruit, all the Trees in the Island having been blown up by the Roots." Much of Jamaica was in a similar state of crisis after the Savanna-La-Mar Hurricane, and Jamaica's problems were compounded by still another powerful hurricane in 1781, which left "not a plantain tree hardly standing nor any ground provisions of any sort" in some parts of Jamaica. As is often the case during a famine, diseases compounded the death toll, especially among the slaves, thanks to widespread malnutrition as well as the contamination of water sources. Worst of all, the region's starving slaves were saddled with almost the entire burden of post-storm rebuilding in addition to their normal plantation tasks. Small wonder, then, that 15,000 slaves or more may have succumbed to the hurricanes of 1780 and their aftermath in the British Caribbean. Incidentally, this high death toll pushed planters toward the same choice that the Taíno had made centuries before: to switch over to root crops rather than orchards or grains to feed their slaves, since root crops were far more resilient against hurricane winds.

The 1780 hurricanes might have exerted an even higher death toll if not for large-scale British relief aid in the aftermath of the storm. Once the misfortunes of the British Caribbean became known to the metropole, the government opened its coffers and private individuals opened their purses, contributing heavily to relief funds for the region. To some degree, this generosity reflects the tenor of the times. Mulcahy notes that the late-eighteenth century has been called an "age of benevolence," during which it was fashionable to feel empathy and show compassion for one's fellow human beings. However, political considerations may have played an even greater role in encouraging generous humanitarian relief. Mulcahy notes that, given the American fight for independence and the ongoing war with France and Spain, "disaster relief … made good political sense," as it "reaffirm[ed] the connection between Britain and the West Indian colonies." A disproportionate share of this relief money would ultimately go to wealthy creole elites rather than the poorest and neediest inhabitants of Britain's colonies, which further reinforces the notion that the post-hurricane aid served political as much as humanitarian purposes.

While the hurricanes of the eighteenth century would ultimately reinforce British control over its possessions in the West Indies, several authors have contended that the same hurricanes may have played a sharply different role in British North America, helping to securing American independence. According to Johnson (2011), hurricanes in the 1770s had triggered repeated famines in the Spanish Caribbean, but the Spanish Empire's rigid mercantilist laws had prevented local creole elites from importing foodstuffs from nearby North American producers to make up for harvest shortfalls. As a result, when faced by recurrent food shortages, Spain began to relax its trade restrictions and increasingly green-lighted trade with English North America. By the mid-1770s, in fact, Spain was becoming reliant on Philadelphia grain merchants to feed its Caribbean empire. This burgeoning market in the Spanish Caribbean, Johnson contends, convinced many American merchants that "they no longer needed Great Britian for their economic survival," and as partly as a consequence, "American Patriots were willing to gamble on independence from Great Britain" in 1776.

In addition, once the conflict was underway, the extraordinary 1780 hurricane season may have altered its outcome by changing the balance of power at sea. The British lost 15 major warships to the hurricanes of 1780, which is more than they lost in all the naval engagements of the Revolutionary war combined. Many more British vessels were dismasted and rendered *hors de combat* by the storm (Figure 6.2), undermining Britain's naval supremacy and foiling the British plan to force the rebellious North American colonies into submission by imposing a smothering naval blockade. What is more, according to Eric Dolin (2020), the hurricanes of

Figure 6.2 Two British ships dismasted and foundering in the 1780 Great Caribbean Cyclone
Source: Fighting Ships, 1765–1783, © Royal Greenwich Museum.

this period convinced the French to play a more direct role in the war with Britain. While France had allied with the American revolutionaries in 1778, the French were initially hesitant to offer direct naval assistance to the fledgling United States out of respect for British naval superiority. After 1780, however, the French became more willing to send their ships north to the mainland North American coast. In part, this was due to British naval losses, but also because doing so offered French admirals the prospect of getting their ships out of dangerous Caribbean waters during the hurricane season. The result was the decisive Battle of Yorktown in September of 1781, where George Washington's continental army joined with a French naval expedition to force the surrender of British General Cornwallis's expeditionary army in the southern colonies. The Revolutionary War would drag on for two more years after Yorktown, but in retrospect at least, Yorktown made American independence all but inevitable.

The Great Bhola Cyclone, 1970

Two centuries later and on the other side of the world, another powerful cyclone would play a similar role in helping to midwife an independent nation. However, as we will see in this section, the birth of Bangladesh was a far more traumatic event than the independence movement that created the United States. Indeed, the Great Bhola Cyclone would prove to be the deadliest cyclonic storm in recorded history, and the post-cyclone war of

Bangladeshi independence was deadlier still. In total, half a million people perished during the cyclone. Another 500,000–3,000,000 would die during the subsequent independence war, in part because the war perpetuated the hunger and deprivation that the cyclone had left in its wake – and perhaps also because post-cyclonic disease cues may have tipped existing animosities into outright xenophobia and massacre.

Building Vulnerability on the Bay of Bengal

The Great Bhola Cyclone of 1970 was deadly in large part because of where it made landfall, the low-lying but densely populated Ganges Delta. This delta, the largest such landform in the world, consists of sediment deposits up to 16 kilometers thick laid down over the course of the past 30 million years by the Brahmaputra, Ganges, and Meghna Rivers. All three of these watercourses are sediment-rich "braided rivers," meaning that they are constantly shifting their courses and creating sub-channels as the steady build-up of alluvial deposits in their riverbeds forces them to find new paths of least resistance to the sea. The cumulative result of this process is an extremely flat and low-lying flood plain, consisting of numerous islands known locally as *char*. Since they consist primarily of sediment deposits, these *char* are ephemeral; the average lifespan of a *char* is only about 4 years and 90% of *chars* succumb to erosion within 20 years of their creation. While they last, however, the *char* can support rich harvests, and thanks to their fertility the Ganges Delta boasted a population density of nearly 400 people per square kilometer in 1970, over four times the population density of contemporary France.

Unfortunately for its many inhabitants, the Ganges delta's low-lying topography and unstable soils render it highly vulnerable to freshwater flooding. The Brahmaputra, Ganges, and Meghna Rivers receive much of the meltwater of the glaciers and snowfields of the Himalaya Mountains, so abnormally heavy floods often follow in the wake of anomalously high and sustained regional temperatures. Catastrophic floods also occur during years of heavy monsoon rainfall in northern India, especially when river crests coincide with each other and with heavy rainfall within the Ganges Delta itself. Such was the case in 1998, when floodwaters covered fully 75% of Bangladesh, displaced 30 million people, and killed an estimated 1,000 Bangladeshis, many of whom succumbed to cholera and typhoid outbreaks that followed in the wake of the floods.

Saltwater flooding in the form of cyclonic storm surge poses an even greater hazard to the inhabitants of the Ganges Delta, especially in the *char* islands surrounding the river's mouth. Bangladesh has the misfortune of lying at the top end of the Bay of Bengal, a sea that routinely boasts temperatures of 26°C or higher, close to the threshold required for cyclone development. What is more, when cyclones do occur, their storm surge is worsened by local topography, since the Bay of Bengal is shaped like a northwards-pointing funnel aimed directly at the *char* islands. As a result, catastrophic storm surges have made frequent and deadly appearances in the pages of the Ganges Delta's history. In 1737, for example, a "super cyclone" inundated the delta under 10–13 meters of storm surge and some contemporaries guessed that the death toll may have reached 300,000, though it is impossible to sure. A similar storm struck the British-controlled Ganges Delta in 1876, and though it only killed 100,000 people directly, it contributed to post-disaster cholera outbreaks that claimed another 100,000 victims. More recently, a 1942 cyclone that struck Bengal during World War II killed at least 42,000 more Bengalis directly and contributed to the deaths of 3 million more in the following year, when famine broke out in Bengal in part due to crop losses during the 1942 cyclone.

Over time, Ganges Delta farmers had developed a repertoire of cultural techniques to mitigate the risks posed by such floods. Farmers typically grew multiple different types of crops (including some root crops) on two or more unconnected strips of land, ideally at different elevations above water level and with different types of soil. The aim of such practices was not to maximize yields, but to minimize the chance of a total harvest failure. Farmers also sought to shield themselves from floods by digging drainage channels around their homes and using the excavated soil to raise their family compounds, or *bari*, above the flood level. These *bari* tended to be surrounded by a dense thicket of coconut palms, water chestnuts, bananas, and other plants, which provided not only food and fuel, but protection for the *bari* compound against cyclonic winds.

Ganges Delta farmers also dealt with the challenges posed by their unique environment through mobility and migration. Some islands that were particularly vulnerable to floods might only be seasonally occupied. If a *bari* was located on a *char* island that was shrinking due to erosion, its component houses could be dismantled and hauled to more stable terrain, or else crucial structural components, like the expensive corrugated metal roofs, could be salvaged and reused. As a last resort, farmers could abandon the countryside entirely and migrate to urban centers, and as we saw in Chapter 5, such migrations are becoming increasingly frequent in the age of global warming. It should be noted that the migration of the delta's farmers was driven, not just by climate adversity, but also by population pressure, since Bangladesh's population was growing in the late 1960s at a rate of up to 3% per year. As a result, in the years before 1970, land-hungry farmers had begun to colonize the coastal Sunderban Forest, one of the largest expanses of mangroves in the world. This migration, Alfred Sommer and W. Henry Mosley (1973) pointed out, was transforming a region that was "once the home of the Bengal Tiger" into a "relatively prosperous rice surplus area, but one constantly exposed to the vagaries of the Bay of Bengal." Unfortunately, the conversion of mangrove swamps to rice paddies along the south coast also served to whittle away the protective seawall of coastal mangroves that helped to protect the interior *char* islands from the worst of the cyclonic storm surge.

The underlying geography of the Ganges Delta, therefore, played an important role in setting the stage for a cyclone disaster. Shorter-term political events also increased the region's vulnerability. At the time of the cyclone, the Ganges Delta was under the sovereignty of Pakistan, a composite nation consisting of two territorially disconnected regions separated by about 2,000 kilometers but united by a common Islamic faith. Unlike West Pakistan, where most citizens spoke Urdu, East Pakistan was ethnically Bengali, and the Bengalis had reacted angrily when politicians in the west of the country had tried to impose Urdu as the national language. The East also resented the fact that western Pakistanis dominated the armed forces, a legacy of British imperialism, since the British had considered the Bengalis to be a "non-martial" race and had sought their recruits primarily from the Punjabi Muslims of the west. Tensions between East and West were also stoked by the fact that the West received the lion's share of state revenue despite the East's larger overall population. As a result of these inequities, Bengali politicians were beginning to call for greater autonomy within the state of Pakistan, though at this point few in the East were yet advocating for outright secession from the Pakistani state. Still, tensions were mounting on the eve of the cyclone, in part because of heated political campaigning in advance of a planned December election.

The timing of the 1970 cyclone was unfortunate in other ways as well. The cyclone struck the delta at high tide during a full moon, greatly exacerbating the height of its storm surge. It also struck in the middle of the night, when most people were asleep in their beds, and the middle of November, a month not usually associated with cyclonic weather. November was also the season of rice harvesting, which was doubly unfortunate: not only were the existing food stocks fairly low by this point in the agricultural calendar, the harvest season had attracted a half-million or more migrant workers from the northern parts of the Ganges Delta, increasing the number of potential victims of the cyclone. Many of these migrant farmers routinely camped out in the open beside the fields that they were paid to harvest and thus had little protection against cyclonic winds and storm surge. Finally, since 1970 was a period of considerable India-Pakistani tension, the meteorological agencies of these feuding nations did little to share information about the oncoming storm, contributing to the region's unpreparedness.

"The Worst Kind of Infidelity": Cyclone, Martyrdom, and Bangladeshi Independence

As a result of poor communications, many inhabitants of the Ganges Delta did not learn about the approaching cyclone until it began to shake their walls and rattle their roofs on the evening of November 12. Many local dwellings, which were constructed primarily of bamboo, palm fronds, and thatched grass, began to disintegrate in the face of sustained 210 km/h winds. However, the deadliest hazard posed by the Great Bhola Cyclone was its storm surge, which swamped the southernmost *char* of the Ganges Delta with up to 10 meters of water. Desperate to stay above the raging floodwaters, people scrambled onto the roofs of their houses or climbed to the top of trees, which explains why male adults aged 15–49 were more likely to survive the cyclone, as they had the upper body strength to hold on against the floodwaters. The same group was also most likely to exhibit "cyclone syndrome" after the storm, severe abrasions to the limbs and chest which, according to Sommer and Mosley (1973), betokened the "tenacity with which the survivors had clung to the trees to withstand the buffeting of the waves."

But victims of cyclone syndrome could count themselves lucky; at least they survived the storm, which could not be said of about half the children living in the worst-afflicted districts south of Dhaka. Women and the elderly also died disproportionately during the cyclone, as did the poor, who generally lived in more exposed or flood-prone locations and occupied lower quality housing. The overall death toll was between 300,000 and half a million, and this was likely an undercount, as it was based partly on post-disaster surveys that did not include fatalities among the temporary workers who had been attracted by the harvest. What is more, in the absence of reliable data collection by the Pakistani state, post-disaster surveys were forced to rely on survivor testimony to enumerate the death toll. As a result, surveyors inevitably undercounted the mortality in the worst-afflicted communities, where almost no one had survived to tell the tale.

The Great Bhola Cyclone finally blew itself out on November 13, but the ordeal of the survivors was far from over. The cyclone struck during what the French called the *soudure*, the hungry period when the previous year's harvest was nearly exhausted and the next harvest was still unripe. As a result, the food stocks in the region were already low and the population was already malnourished even before the cyclone destroyed much of the region's stored supplies. Most of the crops in the fields had been ruined by the storm's howling winds and saltwater surge, and the local fishing fleet, which normally provided 80% of the region's protein, was out of commission. Tens of thousands of the survivors were badly injured, with broken bones and infected wounds, and fears were rising that cholera might exploit the post-storm misery and magnify the death toll, as had happened back in 1876.

Quick government action was therefore needed to prevent a bad situation from becoming even worse. But the West Pakistani-dominated central government seemed oblivious to the suffering of the East. President Yahya Khan reportedly flew over the delta in a military aircraft while on the way back from a trip to China and then, once he returned to the ground, declared that the damage to the Ganges Delta had been exaggerated.

While some Pakistani military resources were mobilized to reach the victims of disasters and distribute supplies, the pace of relief efforts was so slow that Bengali politicians began to accuse the Khan regime of "gross neglect" and "utter indifference." Pakistani authorities also refused to move the date of the scheduled December election, which, given the scale of destruction in the Ganges Delta, provoked charges that the West Pakistani-dominated government was deliberately trying to suppress the vote in East Pakistan.

In the face of a rising crescendo of condemnation, Khan did belatedly return to East Pakistan to take charge of relief operations on November 24, but by then the political damage had been done. The main beneficiary of Khan's plummeting popularity was opposition leader Sheikh Mujibur Rahman, leader of the Bengali nationalist Awami League. Before the cyclone, Mujib had campaigned on greater Bengali autonomy within the state of Pakistan. After the cyclone, however, he sharpened his attacks on the West Pakistani "central government and bureaucrats," whom he accused of "criminal neglect and discrimination against Bangladesh, which has made us so vulnerable a prey to every vagary of nature." According to John P. Thorp (1987), Mujib went so far as to accuse the Pakistani government of the "worst kind of infidelity": of killing fellow Muslims, presumably in an attempt to weaken the Bengali people and thus perpetuate West Pakistani domination over the government. In Mujib's rhetoric, the victims of the cyclone therefore became the "million martyrs," whose blood was on the hands of Pakistani authorities. He therefore argued that the Bengali people must form a new state, Bangladesh, "if they were to escape annihilation" at the hands of the infidel Pakistani elites.

Mujib's post-cyclone rhetorical shift begs the question: did the ubiquitous disease cues that he observed while campaigning in post-storm Bangladesh predispose him toward us-vs-them rhetoric and out-group exclusion through the workings of the behavioral immune system? Possibly, but I don't think that psychoanalyzing single individuals like Mujib is an appropriate task for a historian. While the BIS certainly influences human behavior on a biological level, there is considerable biological variation between individuals, and in any case, any given person's beliefs and actions are the result of numerous influences and multiple motivations. In Mujib's case, for example, it is plausible that his shift toward nationalistic, xenophobic rhetoric was a conscious and calculated rhetorical ploy rather than the unconscious result of a mental process activated by exposure to death and injury. However, we do know from laboratory experiments that, if enough people are exposed to disease cues, the exposed groups will show predictable and statistically significant behavioral changes, specifically heightened in-group preference and out-group animus. If the activation of the BIS after the storm influenced human behavior in Bengal, therefore, we should seek those changes not in single individuals like Mujib, but rather at the level of the aggregate.

So did the Great Bhola Cyclone trigger behavioral changes in the group level in East Pakistan? The results of the 1970 Pakistani election suggests that it did. Perhaps because Mujib's us-vs.-them rhetoric struck a chord in post-cyclone Bengal, the elections led to an overwhelming victory for the Awami League, which won 167 out of 169 seats reserved for East Pakistani delegates in the National Assembly. Mujib had expected to do well in the election, but even he was astounded to find that his electoral success in the East combined with East Pakistan's greater population meant that the Awami League now had more seats in the National Assembly than all other parties combined, leading to the bizarre situation where a successionist party had seized majority control over a national legislature. Unable to accept the Awami League's victory, which made Bangladeshi independence a virtual certainty, Yahya Khan delayed the opening of the National Assembly, started to build up troops in East Pakistan, and tried to negotiate a power-sharing arrangement with Mujib. But these talks were fruitless, in part because Mujib insisted that Bangladesh's claim to nationhood had already been secured by the blood of the million martyrs. Eventually, Mujib unilaterally declared independence, Yahya Khan retaliated by outlawing the Awami League, arresting Mujib, and ordering Pakistani troops to snuff out rebellion in Bengal.

The resulting conflict has been called a civil war, but in reality, it was a pre-planned massacre which some scholars have likened to genocide. The goal of Pakistani troops during "Operation Searchlight," as it was called, was to purge Bangladesh of undesirable elements, including nationalist Bengali politicians, intellectuals, and Hindus. Many Pakistani troops did not need much convincing to commit these atrocities: according to historian R. J. Rummel (1994), West Pakistani officers and soldiers alike had long dehumanized the Bengalis, comparing them to "monkeys and chickens" and agreed with General Niazi's denigration of the region as a "low-lying land of low-lying people." What is more, the Bangladeshi claim that the West Pakistanis were guilty of infidelity found its echo in the West Pakistani belief that Bengalis were "unbelievers," as their religious practices were tainted by Hindu cultural influences. Given such rhetoric, it is not surprising that Hindu residents of Bangladesh were particularly singled out for massacre. In some areas, Pakistani troops killed all uncircumcised males in a community, and Christian Gerlach (2010) reports that some Pakistani military commanders considered the conflict to be the perfect opportunity to liquidate Bengali Hindus, "suspected Indian collaborators" who had "undermined the Muslim masses" and "bled the province white." While these West Pakistani stereotypes about the East long predated the cyclone, it is likely that death, destruction, and other disease cues in post-cyclone Bangladesh helped to shift such anti-Bengali and anti-Hindu biases into outright xenophobia and massacre.

Due to Bengali armed resistance and Indian intervention on Bangladesh's behalf, the Pakistani government was forced to capitulate to

Bangladeshi independence on December 16, 1971. As the symbol of its sovereignty, Bangladesh adopted a simple flag design, a green rectangle (symbolizing Islam, as well as the fertility of the Ganges Delta) marked with a prominent red circle. The disk, which resembles a sun, was intended to be a retort to the moon-shaped crescent on Pakistan's flag. As for the red, it evokes the blood of these killed during Bangladesh's war of independence – and perhaps also the blood of the "million martyrs" who perished during the Great Bhola Cyclone.

Hurricane Katrina, 2005

The Katrina disaster was not supposed to happen. Back in 1965, Hurricane Betsy had swamped New Orleans, killed 58, flooded 164,000 homes, and inflicted over a billion dollars in damage. In response, Congress had passed the Flood Control Act of 1965, which ordered the Army Corps of Engineers to build New Orleans a Hurricane Protection System (HPS) robust enough to protect it from "the most severe meteorological conditions considered reasonably characteristic for that region." By the letter of the law, the Corps should have built a flood control system capable of protecting New Orleans from a category-5 monster such as the Labor Day Hurricane of 1935. But that didn't happen, not even close. Due to financial constraints at both the federal and local level, the Corps eventually constructed a more modest HPS which, they claimed, would protect the city from a category-3 hurricane.

Fast forward to 2005, when Hurricane Katrina passed about 60 miles to the east of the city. While Katrina had briefly reached category-5 status in the Gulf of Mexico, it had dropped to category-3 by the time it touched land in Southern Louisiana, near the marshy mouth of the Mississippi. Katrina's impact on New Orleans was further blunted by the fact that New Orleans was in the front left quadrant of the oncoming storm, where the forward motion of the storm and the circular motion of the winds were moving in opposite directions. As a result, much of New Orleans proper was subjected to only category-1 cyclonic winds and category-1–3 levels of storm surge during Katrina. If the Army Corps of Engineers had done their job, therefore, even their scaled-down HPS should have been enough to shield New Orleans from Betsy-level flooding. At most, New Orleans should have suffered from moderate wind damage and temporary rainwater flooding.

But that didn't happen, not even close. Although New Orleans experienced lower storm surge than its levees had been designed to handle, the HPS suffered numerous, catastrophic failures. The result, according to Gary Rivlin (2015), was "the worst engineering disaster in the world since Chernobyl." Fully 80% of New Orleans was flooded during Katrina, and some low-lying Big Easy neighborhoods were submerged beneath 4 meters or more of seawater. Worst of all, even after Katrina had moved on to the north, New Orleans continued to stew in the contaminated floodwaters, since much of the city lies below sea level and the pumping systems were casualties of the storm. As a result, the trauma of America's worst natural disaster of the twenty-first century would drag out for weeks on America's television screens, prompting us to ask: Who sank New Orleans? Why weren't they better prepared? And how could this be happening in America?

Building Vulnerability on the Mighty Mississippi

In seeking answers to these questions, a good starting point was a fateful decision made over a century ago by the Army Corps of Engineers. The Mississippi River, which along with its many tributaries drains fully 41% of the 48 contiguous US states, was an undeniable asset to the United States in terms of both transportation and agricultural production. But the Mississippi was also a problem. In its natural form, the Mississippi (like the Brahmaputra, Ganges, and Meghna) was a braided river, prone to frequent flooding as sediment deposits raised the riverbed. When sediment rose to critical levels, the Mississippi was subject to sudden "avulsions," or the carving of new channels, which endangered the lives and properties of those living on its banks. To mitigate these threats, the Corps adopted a "levees-only policy" in 1885, with the goal of making the Mississippi scour its own bed of sediment by containing it within a narrow straitjacket of high levees to increase the rate of its flow. This policy would have the added advantage of allowing the fertile floodplains on the banks of the river to be exploited by capitalist farmers using Black sharecropper labor.

The main disadvantage of the levees-only policy is that it didn't work. Since the Mississippi refused to do the job of the Corps and scour itself, sediment continued to build up between the levees, to the point that by the end of the twentieth century the lower Mississippi had become a "hanging river," raised above the surrounding plains like a "large vein on the back of the hand" in the evocative phrase of science writer John McPhee (1989). Any sediment that was not retained in the riverbed was discharged, not into the river floodplains or delta, but into the Gulf of Mexico. As a result, the Mississippi Delta was no longer replenished by new river alluvium from upstream and began to erode and subside, a process hastened by the death of the coastal marshes that depended on river sediment as part of their natural lifecycle.

The main loser in all of this was the city of New Orleans, whose fortunes are inextricably tied to the Mississippi River. The irony of New Orleans is that there were good reasons *for* the city to exist, but no good location *where* it could exist: as geographer Peirce F. Lewis points out, "the Mississippi River demands a city at its mouth but fails to provide any place for one" (Rivlin, 2015). The best of many bad options was a natural ridge of slightly elevated riverside land about 160 kilometers from the mouth of the Mississippi, which was also about as far upriver as ocean-going sailing ships of the eighteenth century could venture. Almost as soon as New Orleans was founded, this "sliver by the river" became a bustling transportation hub, the interface between the inland waterways of the Mississippi River and the international trade networks of the Gulf of Mexico and beyond. In the process, New Orleans began to spill out of the higher elevation riverfront sliver into lower-elevation lands nearby.

New Orleans' urban sprawl sped up considerably during the 1930s and 1940s, when a constellation of factors contributed to the formation of new, lower-lying neighborhoods on the periphery of the original city. In the late 1920s and early 1930s, oil was discovered in the nearby Plaquemines and St. Bernard parishes, pumping wealth into New Orleans. New Orleans expanded further during World War II, when it received significant federal investments with the aim of turning the city into a ship-building center. Construction of a new levee system along Lake Pontchartrain in the 1970s and 1980s further bolstered New Orleans' growth by facilitating its expansion into the previously flood-prone north and east. However, the greatest boost that the federal government gave to New Orlean's growth came in 1968, with the creation of the National Flood Insurance Program. As Andy Horowitz (2010) pointed out, the federal government was now actively encouraging the transformation of marshes into neighborhoods by mitigating the risk to homeowners.

One surprising side-effect of this suburban expansion is that it concentrated some of the white population into lower-lying and flood-prone areas. Because many New Orleans land developers saw "the presence of African Americans" as a "bigger threat than floods to the city's property values," they worked to exclude Blacks from new residential areas they were reclaiming from the marshes. Case in point was St. Bernard's Parish, a low-lying neighborhood constructed east of the city for shipyard workers, which by the 1960 had 24,000 white inhabitants and less than a thousand Blacks. Similarly, the neighborhood of Lakeview in northern New Orleans, settled in the post-war period, was almost exclusively white from the time of its founding. Ironically, the city's African-American population remained concentrated in the relatively higher ground of the old urban core. As Horowitz points out, therefore, New Orleans confounded the predictions of the "theory of environmental racism," which predicts that disaster-prone areas will be disproportionally occupied by disadvantaged minority groups. That being said, many "sliver by the river" neighborhoods that are safely above sea level – such as Algiers, Riverside, Marigny, and the French Quarter – were largely white.

New Orleans' neighborhoods were therefore sharply segregated and were becoming more so over time. According to Niki Dickerson (2010), New Orleans bucked a national trend toward increased urban racial integration at the end of the twentieth century, becoming considerably more segregated from 1990 to 2000. Nonetheless, Black and white neighborhoods had one thing in common: they were sinking. In part, this was due to the natural subsidence of marshland-turned-neighborhood, which settled and compacted after being drained of its water. In the neighborhood of Lakeview, where this problem was particularly acute, some houses sank a meter or more by the turn of the century. Making matters worse, by the early 2000s, the protective marshlands surrounding New Orleans were declining at an astounding rate, losing an area the size of a football field every 38 minutes. Human beings bear a double responsibility for the loss of these protective marshes. Not only had the Mississippi levees built by the Corps starved the delta marshes of new life-giving sediment, oil company engineers had carved hundreds of kilometers of access canals through the freshwater ecosystems south of the city, leading to widespread and deadly saltwater infiltration. As Horowitz (2010) points out, the same oil industry was also contributing to New Orleans' vulnerability through greenhouse gas emissions and resultant global warming, leading to higher seas and more frequent cyclonic storms.

New Orleans' greatest self-inflicted wound, however, was arguably the Mississippi River-Gulf Outlet, a shipping channel better known to locals as MR-GO. Built by the Army Corps of Engineers at the urging of New Orleans' shipping interests, MR-GO moved more dirt than the Panama Canal project, diverting resources away from protective levees and floodgates in the process. It also allowed deadly saltwater flooding to penetrate the marshes directly to the south and east of the city, precisely where it was most vulnerable to hurricane storm surge. By some estimates, MR-GO had contributed to the loss of over 180 square kilometers of marshland (34,000 football fields) by the early 2000s. Perhaps worst of all, according to marine scientist Ivor van Heerden, the levee on the north side of MR-GO combined with the levee on the south side of the Intercoastal Waterway to create a funnel (van Heerden and Bryan, 2006). Like natural funnels such as the Bay of Bengal and the *ria* inlets of the Sanriku coast, this artificial funnel served to focus the force of the floodwaters, putting the low-lying and predominantly African-American neighborhoods of New Orleans East and along the Industrial Canal at risk. And MR-GO was not worth this risk: the canal attracted far too little shipping traffic to justify its enormous construction and maintenance costs.

By the time Katrina struck, in the summarizing words of Horowitz (2010), "New Orleans was effectively on the coast, its residents falling ever deeper into their hydrological hole, separated from the increasingly strong hurricanes in the Gulf of Mexico only by the Army Corps' levee system." But if New Orleanians were putting their faith in the Hurricane Protection System, that faith was misplaced. The Corps was supposed to have finished the HPS by 2008, but due to delays and resource diversions, a decade of work still remained on the project when Katrina struck. What is more, much of the work that had been done had not been done well. Rather than building T-wall levees, which have additional bracing and are better protected from overtopping, the Corps tended to build I-wall levees, which are cheaper but offer less protection from storm surge. During the post-Katrina inquest, researchers discovered that some of these levees had been built upon soil with the "stability of oatmeal," which made as much sense as "putting bricks on Jell-O." In other places, van Heerden claims, the Corps had inexplicably failed to build any levees at all, creating spillways into the city that Katrina exploited in August of 2005.

In fairness, the Corps does not bear sole responsibility for the engineering failures that contributed to the Katrina disaster. New Orleans is drained of rainwater pumping it onto large, roughly parallel drainage canals, which resemble the alley drains of the southern Chinese "comb villages" discussed above. Since over 60% of New Orleans is below sea level, however, each of these drainage canals was a potential vulnerability if the city was threatened by storm surge. Thirty years before Katrina, the Corps had recognized this danger and had advocated building closable floodgates on the mouths of the canal as a countermeasure. But the Corps' proposal had been vetoed by the local Sewerage and Water Board, which wanted to focus on pumping capacity, and the Orleans Levee District Board, which balked at the high annual price tag they would be saddled with for the maintenance of these structures. As a result, these protective structures were never built, contributing greatly to the Katrina disaster.

This geographic vulnerability was compounded by economic vulnerability, stemming from the pervasive poverty of the Big Easy. According to J. R. Elliot and J. Pais (2006), while some parts of the "New South" like Atlanta and Houston experienced impressive growth after the 1960s, New Orleans lagged behind, mired in long-term demographic and economic stagnation. A study by Alan Berube and Natalie Holms in 2015 found that, on the eve of the storm, New Orleans was second only to Fresno, California, in its percentage of residents living in neighborhoods where 40% of the inhabitants had incomes below the national poverty line. Given the city's legacy of enslavement and racial discrimination, it is hardly surprising that most of these impoverished parishes were African-American-majority neighborhoods, including Mid-City, Central City, Gert Town, East New Orleans south of I-10, the Lower Ninth Ward, and above all Desire, which by the early twenty-first century was more notorious for its housing projects than its streetcars. Poor but white, the working-class neighborhood of St. Bernard was a notable exception to this rule.

As Mia Bay (2010) has convincingly demonstrated, one important way in which Black poverty contributed to the Katrina disaster was by creating stark inequalities in car ownership. Fully 24% of African Americans living in New Orleans lacked access to a family car in 2000, as compared to only 7% of whites, a difference Bay blames on greater Black poverty, discrimination against Blacks by car dealerships, high car insurance rates in Black zip codes, and lack of parking in many African-American neighborhoods. Not only did lack of access to cars inhibit Black economic opportunities, it also ensured that Blacks would be disproportionately left behind during Katrina's evacuation, since New Orleans' evacuation plans made the rather racist assumption that all citizens had access to personal transportation. As Bay points out, the city's website on the eve of Katrina gave instructions for automotive evacuation, but "provided no instructions whatsoever for leaving the city via any other means of transportation," other than to make vague promises that "special arrangements will be made to those unable to transport themselves." As we will see in the next section, this passive-voice promise was not kept.

High Water and High Tensions in New Orleans

In some ways, the evacuation of New Orleans was a success. Although many people who took to the highways to escape Katrina got stuck in multi-hour traffic jams, fully 80% of New Orleans' population heeded the mandatory evacuation order on the morning of August 28. For those who stayed, however, Katrina was a nightmarish experience. New Orleanians who were unable to evacuate, who tended to be disproportionately poor, Black, and elderly, were told to go to the New Orleans Superdome for shelter from the storm. After hurrying to the Superdome, however, many of these internal evacuees were forced to wait two hours in a line that stretched 3 kilometers from the stadium just to get through the metal detectors at the security gate. Ironically, once inside the Superdome, the evacuees experienced very little security in the poorly-policed shelter. They also found little food and water and even less comfort amidst the hard plastic benches of the stadium. Then Katrina struck, knocking out the Superdome's power and with it, Superdome's air conditioning and plumbing (Figure 6.3). FEMA employee Marty Bahamonde graphically chronicled what happened next. New Orleans' "shelter of last resort," he remembered, "cascaded into a cesspool of human waste and filth. Hallways and corridors were used as toilets, trash was everywhere, and among it all, children, thousands of them." Conditions were even worse at the nearby Morial Convention Center, which would eventually attract 25,000 displaced refugees despite having no stocks of food, water, or medicine whatsoever.

At least the occupants of the Superdome and Morial were relatively dry. On the morning of the 29th, even before Katrina had reached New Orleans, a section of I-wall levee on the Industrial Canal gave way to the incoming storm surge. This breach was soon joined by further levee failures on the 17th-Street and London drainage canals. Since no flood gates had been built on these canals, there was nothing to stop Lake Pontchartrain from pouring into the empty bowl of the city, sweeping away houses near the wall breaches in the process. In total, researchers would discover that the HPS levees had been breached by Katrina's storm surge in 53 different places, an astounding record of engineering ineptitude. While the

Figure 6.3 Refugees in the New Orleans Superdome, September 2, 2005
Source: Mario Tama/Getty News Images.

"sliver by the river" was untouched by the resulting floods, parts of the Ninth Ward were submerged under 2 meters of water, as was Gert Town and Central City. In white Lakeview and St Bernard, water levels reached 3 meters in depth on average, while the Black neighborhoods of New Orleans East experienced 4-meter floods. Survivors in the worst-affected areas reported that, as water filled their attics, they were forced to chop through their own ceilings to get to the safety of the rooftop above.

In a normal tropical cyclone, the storm surge retreats to sea after the passage of the storm. However, since much of New Orleans was below sea level, the floodwaters had nowhere to go. As a result, in the words of Erik Dolin (2020), New Orleans was awash with "hazmat gumbo" in the days after the storm, since the stagnating floodwaters were tainted by sewage, automotive oil, and innumerable bacteria. Douglas Brinkley (2006) notes that many first responders suffered from "Katrina Cough" – "a bronchial hack that left rescue workers feeling dizzy and heavy-headed" – due to prolonged exposure to the water and its contaminants. The situation was even worse for those trapped by the floodwaters inside their homes or atop their roofs. When food or water ran out, many such survivors braved the toxic waters and waded their way to the already-crowded Superdome or Morial Convention Center. Still others struck out for higher ground, which in much of New Orleans meant the elevated overpasses on interstate I-10 or I-610.

Whether they were marooned within their homes, exposed to the elements on the overpasses, or huddled in the filth of the Superdome, the 200,000–350,000 people who had remained in New Orleans during the storm were badly in need of rescue. But help was slow to arrive, at least at the federal level. At the time of Katrina, the US agency tasked with providing post-disaster relief was the Federal Emergency Management Association (FEMA). Unfortunately for New Orleans, FEMA had been starved of funds during the George W. Bush administration, since many Republicans saw FEMA as an expensive entitlement program that carried out functions better performed by the states. Worse yet, after the 9/11 attacks on the World Trade Center, FEMA had lost its autonomy and had been folded into the newly-created Department of Homeland Security. That agency's director, Bush-appointee Michael Chertoff, had raided FEMA's coffers to fund "black-budget programs and assorted doomsday scenarios," designed to forestall another 9/11-style terrorist attack. In the meantime, in the words of Christopher Cooper and Robert Block (2006), "the natural disaster shop that had been so integral to FEMA's original mission" became a "withered appendage." It didn't help that FEMA at the time was manned by Michael Brown, a lawyer and former commissioner for the International Arabian Horse Association with no previous experience in disaster management.

Nor was there any leadership from the top. At the time of Katrina, Bush had been on vacation at his ranch in Crawford, Texas, for nearly a month. Even after news reports began to show people standing on New Orleans rooftops waving for rescue, Bush insisted on prolonging his vacation for two more days. Then, in a move reminiscent of Yahya Khan's flyover of the Ganges Delta after the Great Bhola Cyclone, Bush winged over New Orleans in his presidential plane on his way to Washington to survey the storm damage from above. When a photograph was taken of him frowning out at New Orleans from the window of Air Force One, it was meant by Bush's PR team to show the president's concern for the storm's victims. For many Americans, however, that iconic photo made Bush seem "detached and uncaring," highlighting his distance from the suffering on the ground. Bush would never recover from the damage that Hurricane Katrina inflicted upon his reputation, just as his own father, George Herbert Walker Bush, never overcame the bad press he received for his handling of Hurricane Andrew in 1992.

Both local and federal assistance to New Orleans was slowed by another factor as well: the proliferation of wild post-storm rumors about the lawless barbarity of New Orleans' mostly-Black storm survivors. As Trymaine Lee (2010) has pointed out, in the days after Katrina, the "city was awash in tales of violence and bloodshed." News reporters who should have known better "dipped their pens in purple ink," in the words of Jed Horne (2006) , and repeated some of the most outlandish tales they heard in their media reports, including fantasies such as "an infant's body found in a trash can, of sharks swimming through flood waters on Canal Street, of hundreds of bodies stashed in the Superdome basement." According to Sheri Fink (2013), who studied the impact of Katrina at

New Orleans' Memorial Hospital, doctors and nurses heard that "the inmates of the prison near Tulane have taken over," or that "looters had used AK-47 assault rifles to commandeer postal vehicles, filling them with stolen goods." One doctor at Memorial reportedly feared to leave any patients behind in the facility since "God knows" what these "crazy Black people who think they've been oppressed for all these years by white people … are going to do to these poor patients who are dying. They can dismember them, they can rape them, they can torture them." Worst of all, New Orleans police chief Eddie Compass "verbally exploded" during a live interview with Oprah Winfrey, repeating wild claims like "we've got babies being raped" in the Superdome. At the time, there were no substantiated reports of any rapes in the Superdome, underage or otherwise. According to Trymaine Lee (2010), these wild rumors were born from "the environment that was produced by the storm," which "brought out what was dormant in people here – the anger and the contempt they felt against African-Americans in community." The likely mechanism by which the environment produced these dehumanizing and exclusionary sentiments, of course, is the behavioral immune system, since disease cues like filth, death, and decay were ubiquitous in post-Katrina New Orleans.

One of the recurring tropes that emerged from the sunken city was that of the Katrina looter, almost invariably Black, who was taking advantage of the disaster to plunder the unguarded city. Admittedly, there was some looting of businesses in post-storm New Orleans, especially for food, water, and other essentials, which is hardly surprising given how woefully unprepared civic authorities were for the Katrina disaster. Nonetheless, Dan Berger (2009) claims that the media misinterpreted and exaggerated these incidents, thus conjuring into existence a weapon-toting "looter class," which "quickly became a dominant actor in the Katrina story – depraved, impoverished, pathological, Black." When New Orleans whites were observed to be raiding New Orleans grocery stores for supplies, Horowitz (2010) noted, the media tended to forgive such transgressions as acts of desperation during an emergency. However, when African-Americans looted the same stores, it was seen as proof positive that Blacks were "devolving into criminal, savage freeloaders." Media expert David Perlmutter, Berger (2009) reports, was so incensed by the flawed news coverage of Katrina that he advocated for the creation of a "truth and reconciliation committee studying the media coverage and assigning blame."

Another recurring rumor that dominated the media coverage after Katrina was the fable that snipers, presumably African-Americans, were targeting first responders in the drowned city. Such rumors may owe their origin in part to the DC sniper attacks of October 2002, just three years before Katrina struck. Since the perpetrators of those nationally televised killings were Black, the idea of dusky-skinned snipers picking off people in New Orleans may have seemed credible. And admittedly, some New Orleanians were firing off their guns in the aftermath of the storm, though in many cases these gunshots were an attempt to attract the attention of desperately needed rescuers.

Thanks to the pervasive rumors of armed looters and anarchist snipers, resources were diverted away from rescue operations and toward quasi-military missions to restore order in the flooded city. Berger argues that "some rescue operations were called off entirely," due to fear of unrest in the Big Easy, "while others increased their weaponry and decreased their boat size to boost the security of those still engaged in rescue missions." by September 1, Horowitz (2010) claims, the rhetoric of civil authorities shifted from saving lives to "tak[ing] the city back." Governor Kathleen Blanco of Louisiana, for example, noted in a press conference that the M-16 rifles of the National Guard were "locked and loaded," and the guardsmen had orders to "shoot to kill" any "hoodlums" they encountered on the streets of New Orleans. In the meantime, FEMA rescue teams temporarily withdrew from New Orleans entirely, abandoning their rescue mission due to an unfounded fear of being shot by the people they were supposed to rescue. None of this is to say that no felonies were committed in post-storm New Orleans. New Orleans was notorious for crime even before Katrina stripped away its veneer of civility, and numerous women reported sexual assaults and rapes in the aftermath of Katrina. Still, it should be said that the National Guardsmen sent to the Superbowl expected to find it under the control of Black thugs and piled with corpses, but discovered instead that only six evacuees had died there – four from natural causes, one from a drug overdose, and one from an apparent suicide.

Racist rumors not only slowed evacuation efforts, they got people shot. In Algiers Point on the left bank of the river, Roland Bourgeois Jr. and some of cronies patrolled the streets after Katrina, vowing to keep "outsiders" out of his neighborhood, which in essence meant Black people: he reportedly threatened while on patrol that "anything coming up this street darker than a paper bag is getting shot." Bourgeois made good his promise when he shot 33-year old Donnell Herrington, a Black man whose crime was to stay behind in New Orleans to help his grandparents during the storm. While Herrington survived, others did not, including Henry Glover, an accused looter who was shot dead by New Orleans Police officers and then set afire to hide evidence of the killing. In still another incident, two Black women and four men were shot, and two were killed, on the Danziger Bridge on September 4 by police officers armed with assault rifles. These officers – several of whom were themselves Black – were initially hailed as heroes who had stood up against the forces of anarchy which threatened New Orleans. A subsequent investigation, however, revealed that all victims were unarmed, though the police tried to plant weapons on the scene. What is more, several of the victims, including a developmentally delayed man with the mind of a 6-year-old, had been shot in the back.

Perhaps the most shameful racist incident in the aftermath of Katrina, though it claimed no lives, occurred on the Crescent City Connection

Bridge linking New Orleans to the suburb of Gretna just south of Algiers Point. On September 1, when a mixed group of New Orleans Blacks and ill-starred tourists tried to cross the bridge, their way was blockaded by armed Gretna police officers, who fired over the heads of the oncoming crowd and refused them passage. With nowhere else to go, many of these refugees erected crude shelters and spent the night on the bridge. They hoped to be allowed across the next morning, but their pleas were answered by a police helicopter, which was sent to hover over the encampment, "its rotor blades blowing the makeshift shelters away and pelting people with dust and debris kicked up by the downdraft." Thus, in the words of Douglas Brinkley (2006), Gretna's officials "condemned their fellow humans, of whatever color, back into the unlivable conditions of New Orleans." Gretna officials would later excuse their actions by claiming that a local mall had been robbed and burned the night before. However, this attempt to establish guilt by association only underscores the fact that Gretna officials saw the crowd on the bridge as looters and thugs rather than fellow Americans badly in need of assistance.

Post-Katrina Reconstruction in an Age of Global Warming

Restoring some semblance of normalcy to New Orleans proved to be a long process. The Superdome was not evacuated until September 3, and Morial was only evacuated on September 4. These refugees, along with those plucked by flat-bottomed boats from New Orleans' rooftops and overpasses, were bussed to "temporary" housing in nearby cities like Baton Rouge and Houston. The real turning point for the recovery of New Orleans occurred on September 5, when the gaping hole of the 17th Street Canal breach was finally plugged up. As a result, New Orleans' pumps could finally begin to drain the city of its polluting waters without fear that those waters would pour right back in.

Once the streets of New Orleans had dried, governmental officials began the grim task of counting the enormous costs of the storm and flood. Katrina may have done as much as $180–190 billion dollars in damage to the Gulf Coast, and much of it that economic loss occurred in New Orleans, 80% of which was flooded during Katrina. In total, 1,500 bodies were collected by searchers from city streets, hospital basements, flooded houses, and nursing homes, and the vast majority of these deaths occurred in the greater New Orleans area. Over half of Katrina's victims were Black, and over 70% were elderly or disabled. Many elderly or ill victims of the storm perished due to power outages, which disabled life-saving equipment like respirators and patient monitors. Other elderly New Orleanians died of heat exhaustion after air conditioning failed or did not survive the traumatic process of being transported out of storm-damaged facilities. Katrina's death toll therefore strongly supports Ryan Hagan's (2021) contention that much of the mortality attributable to modern-day disasters in developed nations result from the breakdown of the technological systems upon which society now depends.

The Katrina disaster is also a case in point example of Naomi Klein's (2008) concept of the "shock doctrine" – the idea that elites often exploit natural disasters to force though major structural changes. In the aftermath of Katrina, for example, New Orleans city officials managed to close Charity Hospital, which had served mainly the most underprivileged and needy New Orleanians. City and state officials also forced through major changes to New Orlean's school system, creating a network of charter schools to improve student scores – and break the back of the powerful New Orleans Teachers Union. They also dismantled many of New Orleans' housing projects, though some of them had suffered little damage during the flooding. New Orleans elites also tried to re-make the city by converting some of the lowest-lying districts to parklands or marshes. However, since many of the districts that planners considered evacuating were African-American neighborhood, critics accused the city planners of racism. Barney Frank, a Massachusetts Congressman, compared the plans to "ethnic cleansing by inaction."

In the face of this backlash, the so-called "parks project" was ultimately scuttled, but even so, New Orleans is a much whiter city now than it was at the time of Katrina. In the aftermath of the storm, much of the city was bought up by middle-class "YURPS" – young urban rebuilding professionals – who were largely white and mostly not native New Orleanians. As a result, New Orleans is now one of the most rapidly gentrifying cities in America, and rising rents have made it difficult for many Black former residents to return to their native city. Substantial communities of Black New Orleans refugees still exist in Dallas, Atlanta, Memphis, and above all Houston, where they number 30,000 or more.

Despite the work of the YURPS, or perhaps in part because of it, New Orleans is also a smaller city today than before the storm. While New Orleans proper had nearly half a million inhabitants in 2000, as of 2020, the population had fallen to under 400,000. In the mind of some critics, however, even 400,000 is far too high, given the city's geographical and hydrological problems. As early as 1975, Louisiana biologist and environmental activist Michael Tritico had predicted that "In the long haul, New Orleans, as we know it, will have to be abandoned" (Horowitz, 2010) and he advised New Orleanians that it was better to put up with the "hassle of moving than the nightmare of being in this city when a levee does not function properly." In the aftermath of Katrina, Tritico's words seemed prophetic.

Rather than take his wise advice, however, both local and federal authorities doubled down, acting as if New Orleans is simply too big to fail. In the years after Katrina, the Army Corps of Engineers spent $14.5 billion on new flood control structures in New Orleans: new pumping stations, several hundred miles of reinforced dikes and levees, and (belatedly) massive flood gates to seal off drainage canals and waterways in advance of an incoming storm. What is more, MR-GO was closed, though the marshlands it destroyed were not restored. But any respite that New Orleans has gained from these measures is only temporary. The new, heavier levees built after Katrina are already sinking into the region's soft soil. What is more, in 2019, the Army Corps of Engineers admitted that much more work needs to be done to protect the city, as the "consolidation, settlement, [and] subsidence" of the Mississippi Delta's soils, combined with wetland loss and the rise of sea levels, mean that the "risk to life and property in Greater New Orleans will progressively increase." In the meantime, according to Kerry Emanuel (2013), hurricanes are likely to become more commonplace in the Gulf of Mexico as global warming routinely pushes ocean temperatures above the threshold that can support cyclonic development. New Orleans can therefore be likened to a sandcastle in the path of a rising tide, which can be sustained for a while through vigilance and diligent labor, but will ultimately have to be abandoned to the waves.

Further Reading

Chen, Lincoln C. (Ed.) *Disaster in Bangladesh: Health Crises in a Developing Nation*. New York: Oxford University Press, 1973.

Dauer, Quinn P. "State and Societal Responses to Natural Disasters in Latin American and Caribbean History." *History Compass*, vol. 18, no. 2 (2020), doi:10.1111/hic3.12605.

Davies, Pete. *Inside the Hurricane: Face to Face with Nature's Deadliest Storms.* New York: Henry Holt and Company, 2000.

Emanuel, Kerry. *Divine Wind: The History and Science of Hurricanes.* New York: Oxford University Press, 2005.

Eyerman, Ron. *Is This America? Katrina as Cultural Trauma*. Austin, TX: University of Texas Press, 2015.

Frank, Neil L. and S. A. Husain. "The Deadliest Tropical Cyclone in History?" *Bulletin of the American Meteorological Society*, vol. 52, no. 6 (1971), pp. 438–444.

Frank, Thomas and E&E News. "After a $14-Billion Upgrade, New Orleans' Levees are Sinking." *Scientific American*, April 11, 2019.

Greene, Ronnie. *Shots on the Bridge: Police Violence and Cover-Up in the Wake of Katrina*. Boston, MA: Beacon Press, 2015.

Haque, C. Emdad. *Hazards in a Fickle Environment: Bangladesh*. Boston, MA: Kluwer Academic Publishers, 1997.

Levitt, Jeremy I. and Matthew C. Whitaker. *Hurricane Katrina: America's Unnatural Disaster*. Lincoln, NE: University of Nebraska Press, 2009.

Louisiana State University. "Louisiana's Wetlands Are Being Lost at the Rate of One Football Field Every 38 Minutes." *ScienceDaily*, 4 January 2008. Available at: www.sciencedaily.com/releases/2008/01/080104112955.htm

McGill, Kevin. "Man Gets 10-Year Sentence in Algiers Point, Post-Katrina Racial Shooting." Associated Press, February 14, 2019.

Neely, Wayne. *The Great Hurricane of 1780*. Bloomington, IN: iUniverse, 2012.
Nizamuddin, K. (Ed.) *Disaster in Bangladesh: Selected Readings.* Dhaka, Bangladesh: Graphtone, 2001.
O'Neill, Karen M. "Who Sank New Orleans? How Engineering the River Created Environmental Injustice." In Keith Wailoo, Karen M. O'Neill, Jeffrey Dowd, and Roland Anglin (Eds.), *Katrina's Imprint: Race and Vulnerability in America*. New Brunswick, NJ: Rutgers University Press, 2010, pp. 9–20.
Philbrick, Nathaniel. *In the Hurricane's Eye: The Genius of George Washington and the Victory at Yorktown*. New York: Viking Press, 2018.
Reilly, Benjamin. *Tropical Surge: A History of Ambition and Disaster on the Florida Shore*. Sarasota, FL: Pineapple, 2005.
Reilly, Benjamin. *Disaster and Human History: Case Studies in Nature, Society, and Catastrophe.* 2nd ed. Boone, NC: McFarland, 2022.
Samad, M. A. *Cyclone of 1970 and Agricultural Rehabilitation*. Dacca, Bangladesh: Agricultural Information Service, 1971.
Schwartz, Stuart B. "Hurricanes and the Shaping of Circum-Caribbean Societies." *Florida Historical Quarterly*, vol. 83, no. 4 (2005), pp. 381–409.
Wailoo, Keith, Karen M. O'Neill, Jeffrey Dowd, and Roland Anglin (Eds.) *Katrina's Imprint: Race and Vulnerability in America*. New Brunswick, NJ: Rutgers University Press, 2010.
Walshe, Rory A., Robert M. Rouphail, George C. D. Adamson, and Ilan Kelman. "Werewolves and Warning Signs: Cultural Response to Tropical Cyclones in Mauritius." *Geoforum*, vol. 133 (2022), pp. 56–65.
Webber, Oscar. "The Plantation's Role in Enhancing Hurricane Vulnerability in the Nineteenth-Century British Caribbean." *Alternautas*, vol. 5, no. 2 (2018), pp. 29–42.

Works Cited

Bay, Mia. "Invisible Tethers: Transportation and Discrimination in the Age of Katrina." In Keith Wailoo, Karen M. O'Neill, Jeffrey Dowd, and Roland Anglin (Eds.), *Katrina's Imprint: Race and Vulnerability in America*. New Brunswick, NJ: Rutgers University Press, 2010, pp. 21–33.
Bentzen, Jeanet Sinding. "Acts of God? Religiosity and Natural Disasters Across Subnational World Districts." *The Economic Journal*, vol. 129 (2019), pp. 2295–2321.
Berger, Dan. "Constructing Crime, Framing Disaster: Routines of Criminalization and Crisis in Hurricane Katrina." *Punishment and Society*, vol. 11, no. 4 (2009), pp. 491–510.
Berube, Alan and Natalie Holmes. "Concentrated Poverty in New Orleans Ten Years After Katrina." *Brookings*, August 27, 1015. Available at: https://www.brookings.edu/articles/concentrated-poverty-in-new-orleans-10-years-after-katrina/.
Brinkley, Douglas. *The Great Deluge*. New York: HarperCollins, 2006.
Cooper, Christopher, and Robert Block. *Disaster: Hurricane Katrina and the Failure of Homeland Security*. New York: Henry Holt and Company, 2006.
Cooper, Jago. "Fail to Prepare, Then Prepare to Fail: Rethinking Threat, Vulnerability, and Mitigation in the Precolumbian Caribbean." In Jago Cooper and Payson Sheets (Eds.), *Surviving Sudden Environmental Change*. Boulder, CO: University Press of Colorado, 2012.

Dickerson, Niki T. "The Katrina Diaspora: Dislocation and the Reproduction of Segregation and Employment Inequality." In Keith Wailoo, Karen M. O'Neill, Jeffrey Dowd, and Roland Anglin (Eds.), *Katrina's Imprint: Race and Vulnerability in America.* New Brunswick, NJ: Rutgers University Press, 2010, pp. 169–179.

Dolin, Eric Jay. *A Furious Sky: The Five-Hundred-Year History of America's Hurricanes.* New York: Liveright, 2020.

Elliot, James R. and Jeremy Pais. "Race, Class, and Hurricane Katrina: Social Differences in Human Responses to Disaster." *Social Science Research*, vol. 35 (2006), pp. 295–321.

Emanuel, Kerry. "Downscaling CMIP5 Climate Models Show Increased Tropical Cyclone Activity over the 21st Century." *PNAS*, vol. 110, no. 30 (2013), pp. 12219–12224.

Fink, Sheri. *Five Days at Memorial.* New York: Crown, 2013.

Gerlach, Christian. *Extremely Violent Societies.* Cambridge: Cambridge University Press, 2010.

Hagan, Ryan. "Acts of God, Man, and System: Knowledge, Technology, and the Construction of Danger." In Jacob A. C. Remes and Andy Horowitz (Eds.), *Critical Disaster Studies.* Philadelphia, PA: University of Pennsylvania Press, 2021, pp. 32–50.

Horowitz, Andy. *Katrina: A History, 1915–2015.* Cambridge, MA: Harvard University Press, 2010.

Horne, Jed. *Breach of Faith: Hurricane Katrina and the Near Death of an American City.* New York: Random House, 2006.

Johnson, Sherry. *Climate and Catastrophe in Cuba and the Atlantic World in the Age of Revolution.* Chapel Hill, NC: University of North Carolina Press, 2011.

Klein, Naomi. *The Shock Doctrine: The Rise of Disaster Capitalism.* London: Picador, 2008.

Lee, Trymaine. "Rumor to Fact in Tales of Post-Katrina Violence." *New York Times*, August 26, 2010.

Longshore, David. *Encyclopedia of Hurricanes, Typhoons, and Cyclones.* New York: Checkmark, 2000.

Louie, Kin-sheun and Kam-biu Lieu. "Ancient Records of Typhoons in Chinese Historical Documents." In Richard J. Murnane and Kam-biu Liu (Eds.), *Hurricanes and Typhoons: Past, Present, and Future.* New York: Columbia University Press, 2004, pp. 222–248.

McPhee, John. *The Control of Nature.* New York: Farrar, Straus and Giroux, 1989.

Mulcahy, Matthew. *Hurricanes and Society in the British Greater Caribbean, 1624–1783.* Baltimore, MD: Johns Hopkins University Press, 2006.

Rivlin, Gary. *Katrina: After the Flood.* New York: Simon & Schuster, 2015.

Rummel, R. J. *Death by Government.* Piscataway, NJ: Transaction Publishers, 1994.

Schwartz, Stuart B. *Sea of Storms: A History of Hurricanes in the Greater Caribbean from Colombus to Katrina.* Princeton, NJ: Princeton University Press, 2015.

Sobel, Adam. *Storm Surge: Hurricane Sandy, Our Changing Climate, and Extreme Weather of the Past and Future.* New York: Harper Wave, 2014.

Sommer, Alfred and W. Henry Mosley. "The Cyclone: Medical Assessment and Determination of Relief and Rehabilitation Needs." In Lincoln C. Chen (Ed.),

Disaster in Bangladesh: Health Crises in a Developing Nation. New York: Oxford University Press, 1973, pp. 119–132.

Studholme, Joshua *et al.* "Poleward Expansion of Tropical Cyclone Latitudes in Warming Climates." *Nature Geoscience*, vol. 15 (2022), pp. 14–28.

Tao, Jin *et al.* "Climate-Adaptive Design of Historic Villages and Dwellings in a Typhoon-Prone Region in Southernmost Mainland China." *International Journal of Architectural Heritage*, vol. 16, no. 1 (2022), pp. 117–135.

Thorp, John P. "Sheikh Mujibar Rahman, a Cyclone and the Emergence of Bangladesh." *South Asia Research*, vol. 7, no. 2 (1987), pp. 143–167.

Van Heerden, Ivor, and Bryan, Mike. *The Storm: What Went Wrong and Why During Hurricane Katrina*. New York: Viking, 2006.

Conclusion

The case studies in the last six chapters have taken us from the ancient world to the Anthropocene, from Eurasia to the Americas, from the outer mantle to the upper atmosphere, and even into the evolved structures of the human brain. So, what have we learned in the process?

One recurring theme of the text, I would contend, is that natural disasters have played an important role in shaping human history, on both short-term and long-term time scales. In the short run, natural disasters have contributed to major political shifts, including the fall of the Ming dynasty, the rise of independent Bangladesh, and the outbreak of both the American and French Revolutions. In the *longue durée*, recurrent natural disasters have helped to shape specific "cultures of disaster," systems of belief that help societies understand and mitigate the hazards of their environment. Historical natural disasters may even help to explain observed cultural variation along the individualism-collectivism and tightness-looseness continuums, both between and within modern states.

Clearly, natural disasters matter. But as we have seen repeatedly throughout the case studies, local contexts matter too. The same natural hazard can play out very differently in different societies due to specific local geographic, socio-economic, cultural, and political factors. These factors, in turn, are rooted in historical processes that might stretch back half a millennium, as was the case in Oliver-Smith's (2020) "five-hundred-year" Peruvian earthquake. Fully understanding any disaster, therefore, requires us to comprehend not only the underlying science of that hazard, but also the deep history of the human vulnerabilities that transform these hazards into disasters.

As we saw repeatedly in this text, the most universal and cross-cultural source of vulnerability is poverty. Whether they are living in East New Orleans, the *char* islands of Bangladesh, the hillside slums of Port-au-Prince, or the Low City of Tokyo, the poor disproportionately inhabit landscapes that are vulnerable to disaster. In addition, some impoverished populations, like the Aceh of Northern Sumatra and the plantation slaves of the British Caribbean, were already in a state of crisis or near-crisis even before they were impacted by natural hazards. Such pre-existing

DOI: 10.4324/9781003436805-8

poverty and marginalization not only serve to exacerbate the disaster, they also render the prospect of "returning to normal" after the disaster meaningless, since normalcy would only re-create the vulnerabilities that tipped the hazard into disaster in the first place. While technical systems like levees and sirens can certainly lower the death toll of a disaster, therefore, poverty reduction offers the most effective comprehensive remedy against the overall impact of natural disasters.

Another cross-cultural tendency that we saw repeatedly throughout this text was out-group hostility and xenophobia in the aftermath of natural disasters, occasionally to the point of violence and genocide. Disasters create scapegoats, who might be weather-working witches, well-poisoning Koreans, grain-hoarding Jews, baby-raping New Orleans Blacks, scheming French aristocrats, or eyeball-stealing western missionaries and their "rice Christian" accomplices. As I have argued repeatedly in this text, the best explanation for this seemingly universal tendency is the human behavioral immune system, which is triggered by the hunger and disease cues that almost invariably follow large-scale disasters. This is not to say that xenophobia and massacre are inevitable after disasters, or that they are rooted in human biology alone. Culture is also crucial to the process: the behavioral immune system triggers hostility to outsiders, but "outsiders" are a cultural construct as much as they are a self-evident fact. And politics plays an important role as well. The strong leadership of the Marquis de Pombal may have forestalled violence against Protestants after the Lisbon Earthquake. Conversely, Mujib explicitly invoked the blood of a "million martyrs" – a potent disease cue – to stir up anti-Pakistani sentiment in the desperate aftermath of the Great Bhola Cyclone.

A final recurring theme in this text is the contrasting ways in which traditional and modern societies experience natural disasters. For most of the historical past, the bulk of the death and destruction during disasters was caused directly by the natural hazards themselves, and this continues to be true in parts of the modern developing world. However, in industrial societies, the impact of natural hazards is felt more indirectly, through the disruption of technical systems upon which modern societies increasingly depend. A case in point was Hurricane Katrina, where the failure of air conditioning and lifesaving equipment during the post-storm blackout accounted for the lion's share of the deaths. What is more, disasters in industrialized societies are increasingly the result of self-inflicted wounds, as epitomized by the meltdown of the Fukushima Daiichi nuclear plant during the Tōhoku Tsunami.

The ultimate self-inflicted wound of the modern age, of course, is the man-made catastrophe of anthropogenic climate change due to greenhouse gas emissions since the Industrial Revolution. If the past is any guide to the future, I suspect that human beings will eventually develop a "culture of disaster" for global warming capable of mitigating some of its worst impacts. Nonetheless, all cultures of disaster come at a cost, and, in this case, that cost will likely be borne mainly by impoverished and marginal populations such as the *char* farmers of coastal Bangladesh and the Nets'aii Gwich'in of central Alaska. The real question surrounding climate change in the Anthropocene is not whether humanity will survive, but whether industrialized nations will accept responsibility for their actions (and emissions) and help to mitigate global warming's deleterious impact on the developing world.

Works Cited

Oliver-Smith, Anthony. "Peru's Five-Hundred-Year Earthquake: Vulnerability in Historical Context." In Anthony Oliver-Smith and Susanna M. Hoffman (Eds.), *The Angry Earth: Disaster in Anthropological Perspective*. New York: Routledge, 2020.

Index